I0834899

BUYERS' GUIDE

AND

MECHANICS' MANUAL,

FOR THE USE OF

RAILWAY OFFICIALS,

CONTAINING A COMPLETE LIST OF ALL ARTICLES WANTED BY

RAILWAY COMPANIES,

WITH THE NAMES OF

EXCLUSIVE FIRST HANDS,

AMONG

Manufacturers, Importers and Commission Merchants

IN EVERY LINE; ALSO, A NUMBER OF

VALUABLE TABLES AND RECIPES.

Published by E. W. BULLINGER, 75 Fulton St., N. Y.

New York.

J. W. PRATT, PRINTER, 75 FULTON STREET.

1878.

S. S. TOWNSEND,

31 Liberty Street, - - New York,

GENERAL AGENT FOR

P. H. & F. M. ROOT'S

PATENT FORCE BLAST ROTRY BLOWERS.

BLACKSMITHS' HAND BLOWERS,

AND

IMPROVED PORTABLE FORGES.

Address, for Circulars and Price List,

S. S. TOWNSEND,

31 LIBFRTY STREET, NEW YORK.

INDEX TO BUSINESS DIRECTORY.

For Index to Tables, Recipes, &c., see pages 15, 17, and 19.

See Notices at Foot of Page 13.

BOYNTON'S PATENT

LIGHTNING SAWS!

Also, manufacturer of all kinds of first-class Saws and Saw Handles.

The inventor of this Saw claims that its superiority over all others is now established beyond dispute. No other manufacturer has yet ventured to put his Saws in competition with it at the AMERICAN INSTITUTE or elsewhere, and the challenge of $500 for a public trial has nev r been accepted. It is needless, therefore, to speak of the merits or defects of the various Saws now on the market. The intelligent workman will select the tool that will do the most work with the least labor, and which, at the same time possesses the requisite s rength and simplicity of construction, in which respects this Saw has proved itself far ahead of all others, as will be seen by the Testimonials, which will be mailed to applicants.

Awarded Medal of American Institute, over all competitors, Nov. 1872.

These Saws have also been highly spoken of in the following well known papers, copies of which can be seen at the office of the inventor, viz: "Scientific American," "American Artizan," "New York Tribune," "New York World," "Christian Union," "Christian Radical," "Examiner and Chronicle." "American Agriculturist," "Hearth and Home," "Iron Age," "Liberal Christian," "Baptist Union," "Kansas Tribune," "Davenport (Iowa) Gazette," "Davenport Democrat," "Moore's Rural New Yorker," "New York Evangelist," "New York Observer," "Independent," "The Western Rural," "Chicago Standard," and many others.

BOYNTON'S CROSS-CUT SAWS

In quality of material are fully equal to any in the market, a d each blade is subjected to a rigid inspection before leav ng the factory. They are made from 4 to 10 feet in length.

CAUTION.

As these Saws have been largely imitated, (but all omitting some essen ial principal,) purchasers should be very careful to buy none but those having the name and trade mark of E. M. Boynton, with the $500 challange and directions for filing, using, &c., etched on them.

BOYNTON'S ONE-MAN CROSS-CUT SAW

Will be found very suitable for *Carpenters, Bi dge Builders,* and all who work heavy timber. It is, also, well adapted to *Farmer's* use for sawing down trees, cutting up cord-wood, fence posts, and small logs. It can be used by two men whenever necessary, by attaching one of Boynton's Patent Handles, which are removable at pleasure. It will do as much work as four good choppers.

E. M. BOYNTON, Sole Proprietor and Manufacturer,

80 Beekman Street, New York.

ATWATER, WHEELER & CO.

New Haven, Conn.,

IRON MERCHANTS

AND

MANUFACTURERS,

Import through their LIVERPOOL HOUSE, or MANUFACTURE in their NEW HAVEN MILL, all grades of

Merchant Bar Iron,

Sheet Iron,

Scrap Iron,

Old Rails,

Cast Steel,

Spring Steel,

IRON WIRE,

BLACK OR GALVANIZED,

FOR FENCES AND TELEGRAPH LINES.

BIDDLE MANUFACTURING CO.

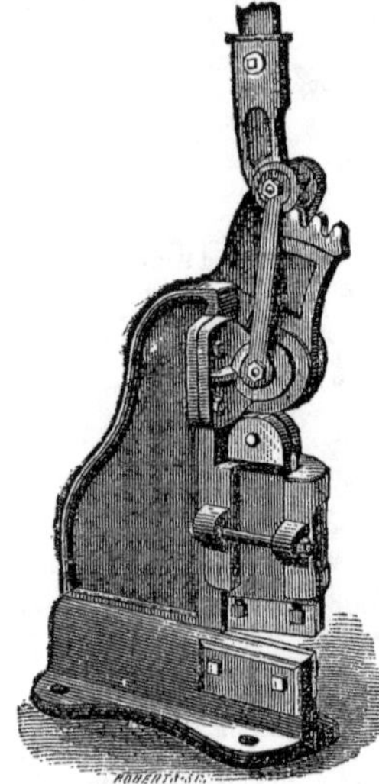

102

Chambers

Street,

NEW YORK

WORKS

AT

Charlotteburg,

NEW JERSEY.

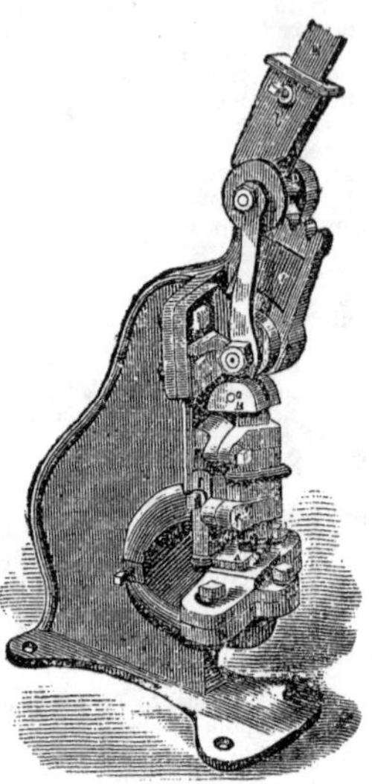

FINE

TOOLS

AND

Hardware

Specialties.

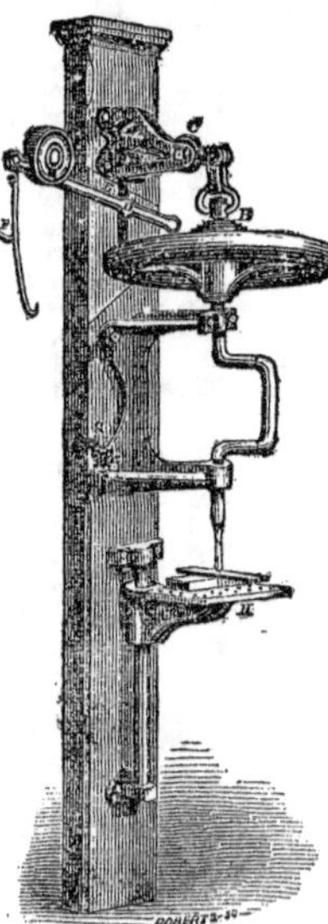

"Eureka" Patent Cutting Nippers.

Combination Wire Cutters and Plyers

Patent Self-Feed Hand Drills,

Hand Shears, Punches, &c.

WITH POWER ATTACHMENT, IF REQUIRED.

The arrangement of the names of the various dealers, under the different heads in the Business Directory, was made by direction of the advertisers, and not at the discretion of the publisher.

Extra copies of this work will be furnished by the Publisher at $1.00 each; address, E. W. BULLINGER, 75 Fulton St., New York

INDEX TO TABLES, RECIPES, &c.

Publisher's Notice.—In selecting the information published herein, I have aimed to give only such tables, &c., as are of practical, every-day use; such as a mechanic will find occasion to refer to very frequently, and by their use, in conjunction with the Silicate Slates (on inside of the covers) and the Memorandum paper, an estimate can be made on almost any kind of an ordinary piece of work, and preserved for future reference.

As will be seen by the heads of the various tables, I have copied from different works published for the use of Mechanics, Engineers, &c., and I beg to advise all who may find the tables herein of value to them, that they buy one or more of these books—they will be found of almost incalculable service; we give below a list of them.

The Civil Engineers' Pocket Book,
by J. C. Trautwine.........Price $5.00, Tucks.

Engineers' and Mechanics' Pocket Book,
by C. H. Haswell............Price $3.00, Tucks.

Mechanics', Machinists' and Engineers' Book of Reference,
by Charles Haslett...........Price $2.50, Tucks.

Useful Information for Railway Men,
by W. G. Hamilton.........Price $2.00.

Useful Formulæ and Memoranda,
by G. L. Molesworth.........Price $2.00.

I will forward any of the above on receipt of the money. Address,

E. W. BULLINGER, 75 Fulton St., New York.

J. G. KNAPP MANUFACTURING CO.,

MANUFACTURERS OF

SCHAEFFER & BUDENBERG, AND BOURDON

STEAM, VACUUM, HYDRAULIC and WATER

GAUGES.

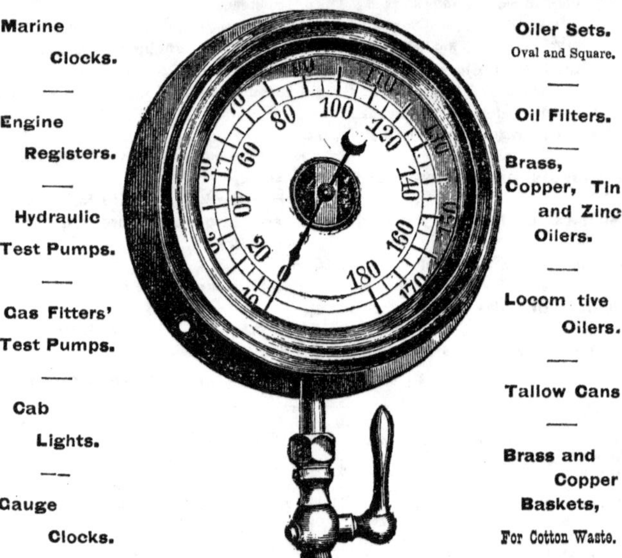

Marine Clocks.

Engine Registers.

Hydraulic Test Pumps.

Gas Fitters' Test Pumps.

Cab Lights.

Gauge Clocks.

Oiler Sets. Oval and Square.

Oil Filters.

Brass, Copper, Tin and Zinc Oilers.

Locom tive Oilers.

Tallow Cans

Brass and Copper Baskets, For Cotton Waste.

ORNAMENTAL BRONZE ENGINE FRONTS,

CONTAINING

Steam and Vacuum Gauge, Counter, Clock and Thermometers,

ALL IN ONE CASE.

Myers' Patent Recording Gauge and Clock.

SOLE MANUFACTURERS IN U. S. OF THE

CELEBRATED GERMAN STUDENT'S LAMPS.

☞ Illustrated Circulars and Photographs sent on application ☜

Office, 26, 28 & 30 Frankfort Street, - NEW YORK.

SILVER MEDAL AWARDED

AT THE

CINCINNATI INDUSTRIAL EXPOSITION,

TO THE

Merchants' Manuf'g and Construction Co.

(TELEGRAPH CONTRACTORS,

PROPRIETORS OF THE

SELDEN PATENT PRINTING TELEGRAPH INSTRUMENT,)

FOR THE

"Best Printing Telegraph Instruments

FOR

PRIVATE USE."

☞ This Company is prepared to build Telegraph Lines for Railroad or other Corporations, or for private Individuals, on Favorable Terms, applying the above Printer, or any other instrument.

W. C. JOY, *President.*
S. J. BURRELL, *Sec'y.*

Merchants' Manufacturing and Construction Co.,
50 BROAD STREET,
P. O. Box 6865. **NEW YORK.**

☞ *SEND FOR CIRCULAR.* ☜

ISAAC U. COLES,

SOLE PROPRIETOR AND MANUFACTURER OF

Patent Tallow Packing for Car Journals,

TALLOW LUBRICATOR AND OIL

FOR MARINE AND STATIONARY ENGINES.

Superintendents, Master Mechanics and Engineers of many prominent Railway Companies, who have tested the **Patent Tallow Packing for Car Journals,** have expressed to me in writing, surprise and satisfaction at its performance, and many of them pronounce it superior to any article heretofore introduced to them for ***Lubricating Purposes,*** and as a RELIABLE PREVENTIVE OF HOT AXLES.

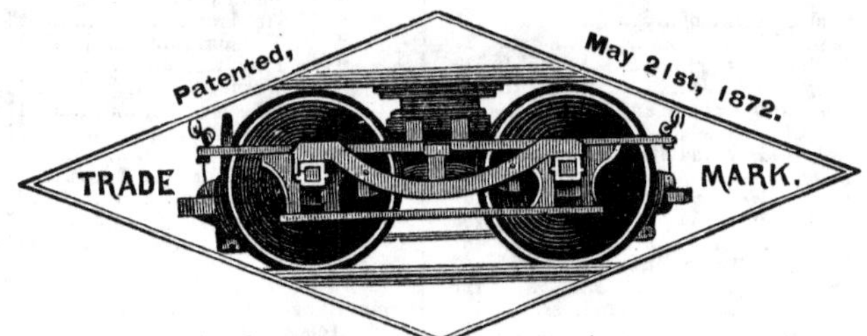

ITS CHARACTERISTICS ARE:

It excludes dust from the journal.
It causes the brasses to wear level.
It can be used in any kind of boxes.
It cannot be spilled out.
It is not affected by difference of climate.
It lubricates more thoroughly.
It requires very little labor and attention.
It is considerably cheaper than any other good lubricating material

The TALLOW LUBRICATOR for Marine and Stationary Engines, and MACHINE OIL, possess advantages over every other similar compound. I wish to secure patronage mainly upon the merits of my materials, therefore, do not here present copies of the many testimonials to be seen at my office from parties who have extensively tested and used the same. My request is "BE YOUR OWN JUDGE."

All Orders for Sample Packages are promptly filled at the office,

20 Cortlandt St., Room 5,

P. O. Box 2963. New York.

BUSINESS DIRECTORY.

Different Articles and Lines of Business in Alphabetical Order, with names of Dealers.

The arrangement of the names, under the various heads in the following Directory, was made by direction of advertisers, and not at the discretion of the publisher.

Adzes.

TEMPERING GRAVERS AND DRILLS.

A watchmaker in Gosling, by the name of Schussleder, has recently published his method of hardening gravers and drills, which he claims renders them almost as hard as the diamond. He first heats the tool to a white heat, and then presses it into a stick of sealing wax, leaves it but a second there, and then sticks it into the wax in another place. This operation is rapidly repeated, until the graver is too cool to enter the wax. Drills tempered in this manner will pierce the hardest steel tempered in the usual manner. In turning or drilling, the tool is moistened with oil of turpentine.

PHILLIPSBURG MANUFACTURING CO.

NEW YORK OFFICE, 23 DEY STREET.

WORKS AT PHILLIPSBURG, N. J.

DANIEL RUNKLE, President.

ALF. P. BOLLER,
Vice-Pres't and Engineer.

WM. RUNKLE,
Treasurer and Sec'y.

JAMES CHRISTIE,
Superintendent.

J. W. GASKILL,
General Agent.

Engineers and Contractors,

FOR THE CONSTRUCTION OF

IRON BRIDGES,

ROOFS, VIADUCTS,

SUSPENSION BRIDGES,

AND

ENGINEERING IRON WORK

IN GENERAL.

ALSO *MANUFACTURERS* OF

SPLICE BOLTS, NUTS,

LIGHT FORGINGS,

BRIDGE, ROOF, and

MACHINE BOLTS.

☞ PLANS, ESTIMATES, &c., FURNISHED ON APPLICATION. ☜

Anti-Friction Metal.

Du Plaine & Reeves....Philadelphia, Pa....See Page 102
G. W. W. Stevens & Co....New York.... " 8
Hook's Smelting Co....Philadelphia, Pa.... " 216
Philadelphia Smelting Co.... " " .... " 134

Anti-Friction Slide Valve.

Geo. W. Rich rdson & Co....Troy, N. Y....See Page 50

Antimony.

Bruce & Cook....New York....See Page 112
Lucius Hart & Co.... " " .... " 56
Philadelphia Smelting Co....Philadelphia, Pa.... " 134
Wm. Green & Co....Wilmington, Del.... " 248

Anti-Vacuum Valve.

Geo. W. Richardson & Co ....Troy, N. Y ....See Page 50

Anvil Blocks.

The Atlas Works....Pittsburgh, Pa....See Page 172

Anvils.

Hermann Boker & Co.... ....New York....See Pages 152, 154
Reese, Graff & Wood....Pittsburgh, Pa....See Page 68

Artesian Well Tubing.

Evans, Dalzell & Co.... ....Pittsburgh, Pa.... ....See Page 58
J. T. Ryerson....Chicago, Ill.... " 180
Morris, Tasker & Co....Philadelphia, Pa.... " 36

Asbestos.

H. W. Johns ....New York....See Page 224

SPEED TABLE FOR TRAINS.

(Molesworth.)

Speed per hour.	Time of Performing.			Speed per hour.	Time of Performing.		
	¼ Mle.	½ Mle.	1 Mle.		¼ Mle.	½ Mle.	1 Mle
Mls.	m. s.	m. s.	m. s.	Mls.	m. s.	m. s.	m. s.
5	3 0	6 0	12 0	33	0 27	0 54	1 49
6	2 30	5 0	10 0	34	0 26	0 53	1 46
7	2 8	4 17	8 34	35	0 25	0 51	1 43
8	1 52	3 45	7 30	36	0 25	0 50	1 40
9	1 40	3 20	6 40	37	0 24	0 48	1 37
10	1 30	3 0	6 0	38	0 23	0 47	1 34
11	1 21	2 43	5 27	39	0 23	0 46	1 32
12	1 15	2 30	5 0	40	0 22	0 45	1 30
13	1 9	2 18	4 37	41	0 21	0 43	1 27
14	1 4	2 8	4 17	42	0 21	0 42	1 25
15	1 0	2 0	4 0	43	0 20	0 41	1 23
16	0 56	1 52	3 35	44	0 20	0 40	1 21
17	0 53	1 46	3 41	45	0 20	0 40	1 20
18	0 50	1 40	3 20	46	0 19	0 39	1 18
19	0 47	1 34	3 9	47	0 19	0 38	1 16
20	0 45	1 30	3 0	48	0 18	0 37	1 15
21	0 42	1 25	2 51	49	0 18	0 36	1 13
22	0 40	1 21	2 43	50	0 18	0 36	1 12
23	0 39	1 18	2 36	51	0 17	0 35	1 10
24	0 37	1 15	2 30	52	0 17	0 34	1 9
25	0 36	1 12	2 24	53	0 17	0 34	1 7
26	0 34	1 9	2 18	54	0 16	0 33	1 6
27	0 33	1 6	2 13	55	0 16	0 32	1 5
28	0 32	1 4	2 8	56	0 16	0 32	1 4
29	0 31	1 2	2 4	57	0 15	0 31	1 3
30	0 30	1 0	2 0	58	0 15	0 31	1 2
31	0 29	0 58	1 56	59	0 15	0 30	1 1
32	0 28	0 56	1 52	60	0 15	0 30	1 0

Asbestos Boiler Felting.

H. W. Johns ....New York......................See Page 224

Asbestos Roofing.

H. W. Johns..New York......................See Page 224

Asphaltum.

H. W. Johns..New York......................See Page 224

Assayers.

Du Plaine & Reeves..................................Philadelphia, Pa............See Page 102

Augur Ratchets.

Barwick Wrench Co...................................Boston, Mass..................See Page 60

Automatic Gear Cutting Machines.

Wm. Sellers & Co.......................................Philadelphia, Pa...........See Page 64

Axe Manufacturers.

G. B. Walbridge..New York......................See Page 130
Geo. Worthington & Co., Wholesale Dealers....Cleveland, Ohio.............. " 100
Vose, Dinsmore & Co...................................New York...................... " 60

Axle Grease.

Frazer Lubricator Co...................................New York......................See Page 264

WATER STATIONS.

The following is taken from Trautwine, and is a fair sample of his comprehensive and valuable method of treating all subjects.

SIZES OF TANKS AND CONTENTS.

Diam.	Depth.	Gallons.	Diam.	Depth.	Gallons.
Ft.	Ft.		Ft.	Ft.	
12	8	6767	24	12	40607
14	9	10363	26	13	51628
16	9	13535	28	14	64481
18	10	19034	30	15	79310
20	10	23499	32	16	96253
22	11	31277	34	17	115451

Cypress or any of the pines answer very well for tanks. The staves may be about 2½ ins. thick for the smaller ones; to 4 or 5 ins for the largest. The bottoms may be the same. The staves should be planed by machinery to suit the curve precisely. Nothing is then needed between the staves to produce tightness. A single wooden dowel is inserted between each two near the top, merely to hold them in place while being put together. The bottom is dowelled together, and simply inserted into a groove very accurately cut, about an inch deep, around the inner circumference of the tub, at a few inches above the bottoms of the staves.

One of 20 feet diameter, and 12 feet deep, may have 9 hoops of good iron, placed several inches nearer together at the bottom of the tank than at the top. Their width 3 inches; the thickness of the lower two, ¼ inch; thence gradually diminishing until the top one is but half as thick. The lower two are driven close together. These dimensions will allow for the rivet-holes for riveting together the overlapping ends; and for a moderate strain in driving the hoops firmly into place. Three rivets of ½ inch diameter, and 3 inches apart, in line, are sufficient for a joint of a lower hoop. One of 34 feet diameter, 17 deep, may have 12 hoops; the lower ones 4 inches by ½; with three ¾ inch rivets to a lower hoop-joint.

The bottom planks of the tank must bear firmly upon their supporting joists, or bearers.

A tank must have a waste pipe for preventing overflow, and a discharge or feed pipe 7 or 8 inches diameter, in or near the bottom. The inner end of the discharge pipe is covered by a valve, to be opened at will by the engine man, by means of an outside cord and lever. To its outer end is generally attached a flexible canvas and gum-elastic hose about 7 or 8 inches diameter and 8 or 10 feet long, through which the water enters the tender-tank. Or, instead of a hose, the feed-pipe may be prolonged by a metallic pipe, or nozzle, sufficiently long to reach the tender; and so jointed as, when not in use, to swing to one side, or to be raised to a vertical position.

Axle Lathes.

Isaac H. Shearman.........Philadelphia, Pa.........See Page 80
Wm. Sellers & Co......... " " .... " 64

Axles.

Anderson & Woods.........Pittsburgh, Pa.........See Page 150
De Laney & Co.........Buffalo. N. Y......... " 98
Greene & Randolph.........New York......... " 54
Jno. A. Griswold & Co.........Troy, N.Y......... " 50
Jones & Laughlins.........Pittsburgh, Pa......... " 32
L. G. Tillotson & Co.........New York......... " 94
Pittsburgh Forge & Iron Co.........Pittsburgh, Pa......... " 174
Vose, Dinsmore & Co. ....New York......... " 60

Babbitt Metal.

A. Fulton's Son & Co.........Pittsburgh, Pa.........See Page 116
Champlin & Rogers.........Chicago, Ill......... " 192
Crane Bros. Mfg. Co......... " " .... " 230
Du Plaine & Reeves.........Philadelphia, Pa......... " 102
J. J. Walworth.........Chicago, Ill......... " 124
Lucius Hart & Co.........New York......... " 56
Phelps & Sanger.........Cleveland, Ohio......... " 120
Philadelphia Smelting Co.........Phila elphia, Pa......... " 134
Wm. Green & Co.........Wilmington, Del......... " 248

WEIGHT OF CAST IRON PIPES.

Weight of a Foot of Cast Iron Pipes, in Pounds.

Two Flanges are equal to One Foot of Length, in Weight.

Bore Inch.	Th ck-ne s of Metal	Weight.	Bore Inch.	Thick-ness of Metal.	Weight.	Bore Inch.	Thick-ness of Metal.	Weight.
1	⅜	5.6	7	⅜	27.3	12	⅝	77.5
1	½	7.4	7	½	36.8	12	¾	94.0
1½	⅜	6.9	7	⅝	46.9	12	⅞	110.8
1½	½	6.8	7	¾	56.9	12	1	127.9
2	⅜	8.7	7½	½	39.3	14	⅝	89.6
2	½	12.0	7½	⅝	50.	14	¾	108 8
2½	⅜	10.5	7½	¾	61.	14	⅞	127.9
2½	½	14 9	8	⅜	30 8	14	1	147.5
3	⅜	1[illegible].4	8	½	41.8	15	¾	115.9
3	½	17.2	8	⅝	52.9	15	⅞	137.
3	⅝	22.2	8	¾	64.5	15	1	157.8
3½	⅜	14.3	8½	½	44.2	15	1⅛	17[illegible].7
3½	½	1[illegible].8	8½	⅝	56.	16	¾	123.
3½	[illegible]	2[illegible].3	8½	¾	68.2	16	⅞	144.7
4	⅜	16.1	9	⅜	34 6	16	1	166.9
4	½	22.1	9	½	46 7	16	1⅛	189 7
4	⅝	28.4	9	⅝	58 9	18	¾	138 9
4½	⅜	18.1	9	¾	71 9	18	⅞	162 8
4½	½	24.5	9½	½	49 1	18	1	18[illegible].8
4½	⅝	31.4	9½	⅝	62.2	18	1⅛	211.8
5	⅜	19.8	9⅜	¾	7[illegible].6	20	⅞	17[illegible].0
5	½	27.0	10	½	51.6	20	1	[illegible]0[illegible].8
5	⅝	34.5	10	⅝	65.3	20	1⅛	233.9
5	¾	42.4	10	¾	79 3	20	1¼	[illegible]61.3
5½	½	29.6	10	⅞	93.5	22	1	227.4
5½	⅝	37.6	10½	½	54.	22	1⅛	255.9
5½	¾	46.	10½	⅝	68.3	22	1¼	28[illegible].8
6	⅜	23.4	10½	¾	[illegible]3.	24	1	245.0
6	½	31.9	11	½	56.5	24	1⅛	277.9
6	⅝	40 7	11	⅝	71 5	24	1¼	311.3
6	¾	49.9	11	¾	86.6			
6½	½	34.4	11	⅞	101 9			
6½	⅝	43.7	11½	½	59.			
6½	¾	53 4	11½	⅝	74.5			
			11½	¾	90.5			

Badges.

L. G. Tillotson & Co. .... New York .... See Page 94
W. W. Wilcox .... Chicago, Ill .... " 190

Baggage Barrows.

A. M. Gilbert & Co. .... Chicago, Ill .... See Page 180
W. A. Dobinson & Co. .... Buffalo, N. Y. .... " 196

Baggage Checks.

L. G. Tillotson & Co. .... New York .... See Page 94
W. W. Wilcox .... Chicago, Ill .... " 190

Baker's Tile.

Philip Neukumet .... Philadelphia, Pa .... See Page 104

Ball Pene Hammers.

Hermann Boker & Co. .... New York .... See Pages 152, 154
Nelson Tool Works .... " .... See Page 30

Band Saws.

Buffalo Machinery Agency .... Buffalo, N. Y. .... See Page 160
C. Van Ness .... New York .... " 70
J. A. Fay & Co. .... Cincinnati, Ohio. .... " 206
J. T. & R. H. Plass .... New York .... " 208
Richards, London & Kelley .... Philadelphia, Pa. .... " 66
Vose, Dinsmore & Co. .... New York .... " 60

Band Sawing Machines.

J. A. Fay & Co. .... Cincinnati, Ohio .... See Page 206
J. T. & R. H. Plass .... New York .... " 208
Richards, London & Kelley .... Philadelphia, Pa. .... " 66

Bar Iron.

See Iron—Bar.

Bark Mills.

R. L. Howard & Son .... Buffalo, N. Y. .... See Page 40

Bar Lead.

Bruce & Cook .... New York .... See Page 112

Bars.

Pittsburgh Bolt Co. .... Pittsburgh, Pa. .... See Page 118

BOLTS WITH SQUARE HEADS AND NUTS.

Weight of 100, *in Pounds.*

Length Inches.	Thickness of Bolt in Inches.							
	¼	5/16	⅜	7/16	½	⅝	¾	⅞
1½	4.16	7.59	10.62	15.94	23.87	39.31		
¾	4.22	7.87	11.72	16.90	25.06	41.38		
2	4.75	8.56	12.38	18.25	26.44	45.69	73.62	
¼	5.34	9.12	12.90	19.38	28.62	49.50	76.	
½	5.97	9.59	14.69	20.69	29.50	51.25	79.75	
¾	6.50	10.44	16.47	21.50	31.16	53.	83.	
3		10.78	17.87	22.38	32.44	56.	85.38	127.25
½		11.81	18.94	26.19	39.75	63.12	93.44	140.56
4			20.59	28.87	42.50	74.87	108.12	148.37
½			21.69	29.87	44.87	79.62	113.12	158.76
5			23.62	32.31	48.81	83.	122.	167.25
½			25.81	34.44	51.38	87.88	128.62	174.88
6			16.87	36.62	53.31	92.38	131.75	204.25
½					56.87	96.88	139.56	214.69
7					59.12	99.87	145.50	228.44
½					61.87	105.75	150.88	235.31
8					64.44	109.50	157.12	239.88
9					70.50	118.12	169.62	258.12
10					77.	128.13	184.	276.18
11					82.88	136.19	195.13	295.69
12					86.37	144.87	209.75	311.94
13					92.	155.50	219.37	335.81
14					97.75	163.58	237.50	351.88
15					103.25	170.75	249.05	391.75

Batteries—Telegraph.

Leclanche Battery Co.......................................New York.................. ....See Page 96
See also, Telegraph Apparatus.

Bellows.

England & Bindley.......................................Pittsburgh, Pa...............See Page 162
Geo. Worthington & Co.......................................Cleveland, Ohio............. " 100
Joseph Churchyard.......................................Buffalo, N. Y.................... " 160
Newcomb Bro's. Sons.......................................New York.................... " 88
Sidney Shepard & Co.......................................Buffalo, N. Y.................... " 196
S. S. Townsend.......................................New York.................... " 2
T. B. Bickerton & Co.......................................Philadelphia, Pa............. " 138

Bells.

A. Fulton's Son & Co.......................................Pittsburgh, Pa...............See Page 116
Vose, Dinsmore & Co.......................................New York.................... " 60

Belt Dogs.

Champlin & Rogers.......................................Chicago, Ill...................See Page 192

Belting.

Bickford, Curtiss & Deming.......................................Buffalo, N. Y................See Page 136
Buffalo Machinery Agency....................................... " " .... " 160
Champlin & Rogers.......................................Chicago, Ill.................... " 192
Darrow, Turner & Co.......................................New York.................... " 62
D. Brewer & Co.......................................Philadelphia, Pa............. " 246
Erie City Iron Works.......................................Erie, Pa.................... " 170
George Worthington & Co.......................................Cleveland, Ohio............. " 100
Goodyears I R. G. Mfg. Co.......................................New York.................... " 252
New York Rubber Co....................................... " .... " 260
N. H. Gardner & Co.......................................Buffalo, N. Y.................... " 254
Radley & McAlister Mfg. Co.......................................New York.................... " 250
Sidney Shepard & Co.......................................Buffalo, N. Y.................... " 196
T. B. Bickerton & Co.......................................Philadelphia, Pa............. " 138
Vose, Dinsmore & Co.......................................New York.................... " 60

Bench Screws—Iron and Wooden Handles.

Hermann Boker & Co.......................................New York...............See Pages 152, 154

THICKNESS OF BOILER IRON REQUIRED AND PRESSURES ALLOWED BY THE LAWS OF THE U. S.

Pressure equivalent to the Standard for a Boiler 42-in. in Diameter and ¼ in Thick.

Thickness in 16ths.	Diameter.						
	34 Inches.	36 Inches.	38 Inches.	40 Inches.	42 Inches.	44 Inches.	46 Inches.
	Lbs.	Lbs.	Lbs.	Lbs.	Lbs.	Lbs.	Lbs.
5	169.9	160.4	152.	144.4	137.5	131.2	125.5
4½	158.5	149.7	141.8	134.7	128.3	122.5	117.2
4¼	147.2	139.1	131.8	125.1	119.2	113.7	108.8
4	135.9	128.3	121.6	115.5	110.	105.	100.
3⅔	124.5	117.6	111.4	105.9	100.8	96.2	92.
3⅓	113.2	106.9	101.3	96.2	91.7	87.5	83.
3	101.9	96.2	91.2	82.6	82.5	78.7	75.

WEIGHT AND THICKNESS OF BOILER IRON.

1-8	inch weighs... 5 pounds per sq. foot.	No. 1 Iron is 5-16 inch thick.
3-16	" " ... 7½ " "	No. 3 " nine-thirty-seconds.
1-4	" " ...10 " "	No. 4 " 1-4 inch.
5-16	" " ...12½ " "	No. 5 " seven-thirty-seconds.
3-8	" " ...15 " "	No. 7 " three-sixteenths.
7-16	" " ..17½ " "	
1-2	" " ...20 " "	

Bending Rolls.

Wm. B. Bement & Son	Philadelphia, Pa	See Page 168
Wm. Sellers & Co	" "	" 64

Bismuth.

Lucius Hart & Co	New York	See Page 56

Black Lead.

T. B. Bickerton & Co	Philadelphia, Pa	See Page 138

Blacksmith's Tools.

Hermann Boker & Co	New York	See Pages 152, 154
Nelson Tool Works	" "	See Page 30

Blank Books.

Allen, Lane & Scott	Philadelphia	See Page 90
Chas. F. Ketcham	New York	" 262
Warren, Johnson & Co	Buffalo, N. Y	" 100

Blasters' Tools.

Geo. W. Mowbray	North Adams, Mass	See Page 182
Nelson Tool Works	New York	" 30
Rand & Waring Drill & Compressor Co	"	See Front Page

Blast Furnaces.

The Atlas Works	Pittsburgh, Pa	See Page 172

Blast Furnace Blocks.

Star Fire Brick Co	Pittsburgh, Pa	See Page 232
Philip Neukumet	Philadelphia, Pa	" 104
Maurer & Weber	New York	" 228

Blasting and Mining Wires.

Bishop Gutta Percha Works	New York	See Page 34
Geo. W. Mowbray	North Adams, Mass	" 182

Blind and Sash Makers.

Joseph Churchyard	Buffalo, N. Y	See Page 160

Blocks.

Providence Tool Co.; H. B. Newhall, Agent	New York	See Page 186

Block Tin.

Bruce & Cook	New York	See Page 112
Lucius Hart & Co	" "	" 56

Blowers.

Wm. B. Bement & Son	Philadelphia, Pa	See Page 168
Phelps & Sanger	Cleveland, Ohio	" 120
S. S. Townsend	New York	" 2

NUMBER OF BURDEN'S RIVETS IN 100 LBS.

Length Inches.	Thickness in Inches. ½	⅝	11/16	¾	Length Inches.	Thickness in Inches. ½	⅝	11/16	¾
¾	1,092	665			3¼	433	267	212	180
⅞	1,027	597			½	413	248	201	169
1	940	538	450		¾	395	241	192	160
⅛	840	512	415		4		230	184	158
¼	797	487	389	356	¼		220	177	150
⅜	760	460	370	329	½		210	171	146
½	730	440	357	280	¾		200	166	138
⅝	711	420	340	271	5		190	161	135
¾	693	390	325	262	¼		180	156	130
⅞	648	375	312	257	½		172	151	124
2	608	360	297	243	¾		164	145	120
⅛	573	354	289	237	6		157	140	115
¼	555	347	280	232	¼		150	138	111
½	525	335	260	220	½		146	134	107
¾	500	312	242	208	¾		143	129	104
3	460	290	224	197	7		140	125	100

THE BISHOP GUTTA PERCHA WORKS,

S. BISHOP,

Nos. 422, 424 & 426 EAST 25th STREET,

NEW YORK.

MANUFACTURE

Submarine Telegraph Cables

AND

Telegraph AND Electric Wires,

ALSO EVERY VARIETY, FOR

BLASTING AND MINING USE.

SILK AND COTTON COVERED WIRES OF SUPERIOR QUALITY

OFFICE WIRES OF EVERY DESCRIPTION.

WATER, BEER, SODA and ACID PIPES.
CHEMICAL VESSELS, FUNNELS and BOTTLES.
CRUDE GUTTA PERCHA for CEMENT.
BATHS and DISHES for ACIDS.
TISSUE SHEET for HATTERS. SURGICAL SHEET.
SHEET FOR ARTISTS AND MECHANICAL USES.

Boiler Covering.

H. W. Johns .... New York .... See Page 224
Van Tuyl Mfg. Co .... " .... " 242

Boiler Felting.

H. W. Johns .... New York .... See Page 224
J. J. Walworth .... Chicago, Ill .... " 124
T. B. Bickerton & Co .... Philadelphia, Pa .... " 138
Van Tuyl Mfg. Co .... New York .... " 242
Vose, Dinsmore & Co .... " " .... " 60

Boiler Iron.

P. H. Moffat .... Chicago, Ill .... See Page 190
Tyng & Co .... New York .... " 260
Vose, Dinsmore & Co .... " " .... " 60
Wm. Green & Co .... Wilmington, Del .... " 248

Boiler Makers.

Erie City Iron Works .... Erie, Pa .... See Page 170
J. B. Waring .... New York .... " 24
Snyder Bros .... Williamsport, Pa .... " 46
Speedwell Iron Works .... New York .... " 122
See also, Steam Boilers.

Boiler Makers' Tools.

Bolen, Crane & Co .... Newark, N. J .... See Page 62
Nelson Tool Works .... New York .... " 30
S. S. Townsend .... " " .... " 2
W. P. Kellogg & Co .... Troy, N. Y .... " 98

Boiler Plate and Drilling Machines.

Thorne, De Haven & Co .... Philadelphia, Pa .... See Page 38

Boiler Plates.

J. T. Ryerson .... Chicago, Ill .... See Page 180
P. H. Moffat .... " " .... " 190
Vose, Dinsmore & Co .... New York .... " 60

Boiler Powders.

H. N. Winans .... New York .... See Page 108

Boiler Ratchets.

Barwick Wrench Co .... Boston, Mass .... See Page 60

AMERICAN AND BIRMINGHAM WIRE GAUGES.

Thickness in Inches.

(Taken from Haswell.)

No. of Gauge.	Thick. of Am. G. Inch.	Thick. of Bir. G. Inch.	No. of Gauge.	Thick. of Am. G. Inch.	Thick. of Bir. G. Inch.	No. of Gauge.	Thick. of Am. G. Inch.	Thick. of Bir. of I ch.
0000	.46	.454	11	.0907	.12	25	.0179	.02
000	.4096	.425	12	.0808	.109	26	.0160	.018
00	.3648	.38	13	.0719	.095	27	.0142	.016
0	.3248	.34	14	.0641	.083	28	.0126	.014
1	.2893	.30	15	.057	.072	29	.0112	.013
2	.2576	.284	16	.0508	.065	30	.01	.012
3	.2294	.259	17	.0452	.058	31	.0089	.01
4	.2043	.238	18	.0403	.049	32	.0079	.009
5	.1819	.22	19	.0359	.042	33	.007	.008
6	.1620	.203	20	.0319	.035	34	.0063	.007
7	.1443	.18	21	.0284	.032	35	.0056	.005
8	.1285	.165	22	.0253	.028	36	.005	.004
9	.1144	.148	23	.0225	.025			
10	.1019	.134	24	.0201	.022			

WROUGHT IRON WELDED TUBES FOR STEAM, GAS, OR WATER.

1¼ inch and below, Butt Welded; proved to 300 Pounds per Square Inch, Hydraulic Pressure.
1½ inch and above, Lap " " 500 " " "

Table of Standard Sizes.—Morris, Tasker & Co.

Inside Diameter.	Actual Inside Diameter.	Actual Outside Diameter	Thickness.	Internal Circumference.	External Circumference.	Length of Pipe per sq. foot of inside surface.	Length of Pipe per sq. foot of outside surface.	Internal Area.	External Area.	Length of Pipe containing one cubic foot.	Weight per foot of length.	No. of threads per inch of screw.
Inches.	Inches.	Inches.	Inches.	Inches.	Inches.	Feet.	Feet.	Inches.	Inches.	Feet.	lbs.	
1/8	.270	.405	.068	.848	1.272	14.15	9.44	.0572	.129	2500.	.243	27
1/4	.364	.54	.088	1.144	1.696	10.50	7.075	.1041	.229	1385.	.422	18
3/8	.494	.675	.091	1.552	2.121	7.67	5.657	.1916	.358	751.5	.561	14
1/2	.623	.84	.109	1.957	2.652	6.13	4.502	.3048	.554	472.4	.845	14
3/4	.824	1.05	.113	2.589	3.299	4.635	3.637	.5333	.866	270.	1.126	11½
1	1.048	1.315	.134	3.292	4.134	3.679	2.903	.8627	1.357	166.9	1.670	11½
1¼	1.380	1.66	.140	4.335	5.215	2.768	2.301	1.496	3.164	96.25	2.258	11½
1½	1.611	1.9	.145	5.061	5.969	2.371	2.01	2.038	2.835	70.65	2.694	11½
2	2.067	2.375	.154	6.494	7.461	1.848	1.611	3.355	4.430	42.36	3.667	8
2½	2.468	2.875	.204	7.754	9.032	1.547	1.328	4.783	6.491	30.11	5.773	8
3	3.067	3.5	.217	9.636	10.996	1.245	1.091	7.388	9.621	19.49	7.547	8
3½	3.548	4.	.226	11.146	12.566	1.077	.955	9.887	12.566	14.56	9.055	8
4	4.026	4.5	.237	12.648	14.137	.949	.849	12.730	15.904	11.31	10.728	8
4½	4.508	5.	.247	14.153	15.708	.848	.765	15.939	19.635	9.03	12.492	8
5	5.045	5.563	.259	15.849	17.475	.757	.629	19.990	24.299	7.20	14.564	8
6	6.065	6.625	.280	19.054	20.813	.63	.577	28.889	34.471	4.98	18.767	8
7	7.023	7.625	.301	22.063	23.954	.544	.505	38.737	45.663	3.72	23.410	8
8	7.982	8.625	.322	25.076	27.096	.478	.444	50.039	58.426	2.88	28.348	8
9	9.001	9.6[illegible]8	.344	28.277	30.433	.425	.394	63.633	73.715	2.26	34.077	8
10	10.019	10.75	.366	31.475	33.772	.381	.355	78.838	90.762	1.80	40.641	8

Boiler Rivets.

Geo. Worthington & Co.	Cleveland, Ohio	See Page 100
Radley & McAlister Mfg. Co.	New York	" 250
Tyng & Co	" "	" 260
Vose, Dinsmore & Co.	" "	" 60
Wm. Gilmor of Wm.	Baltimore, Md.	" 248

See also, Rivets.

Boilers.

H. N. Winans, (Boiler Powders)	New York	See Page 108
J. B. Waring	" "	" 24
Van Tuyl Mfg. Co.	" "	" 242
Wiard Locomotive Attachment Co. (Anti-Explosive)	New York	" 110

See also, Steam Boilers

Boiler Scale Preventive.

H. W. Johns	New York	See Page 224
H. N. Winans	" "	" 108

Boiler Tubes and Flues.

A. Carr	New York	See Page 244
Buffalo Machinery Agency	Buffalo, N. Y.	" 160
Evans, Dalzell & Co	Pittsburgh, Pa.	" 58
J. J. Walworth	Chicago, Ill.	" 124
J. T. Ryerson	" "	" 180
Morris, Tasker & Co	Philadelphia, Pa.	" 36
Pancoast & Maule	" "	" 26
Phelps & Sanger	Cleveland, Ohio	" 120
Tyng & Co	New York	" 260
Vose, Dinsmore & Co.	" "	" 60

WEIGHT OF SHEET AND PLATE IRON.—2 PAGES.

Thickness by Birmingham Wire Gauge and Inches.

Weight of a Square Foot in Pounds.

B. W. Guage.	Thickness, Part of an inch.	Weight, Pounds.	B. W. Guage.	Thickness, Part of an inch.	Weight, Pounds.
36	.004	.162	11	.120	4.88
35	.005	.202		1/8 or .125	5.054
34	.007	.283	10	.134	5.426
33	.008	.322	9	.148	5.98
32	.009	.364		5-32 or .1562	6.305
31	.010	.405	8	.165	6.605
30	.012	.485	7	.180	7.27
29	.013	.526		3/16 or .1875	7.578
28	.014	.595	6	.203	8.005
27	.016	.676		7-32 or .2187	8.79
26	.018	.755	5	.22	8.912
25	.020	.811	4	.238	9.62
24	.022	.912		1/4 or .25	10.09
23	.025	1.018	3	.259	10.37
22	.028	1.137		9-32 or .2812	11.38
	1-32 or .03125	1.259	2	.284	11.525
21	.032	1.31	1	.3	12.15
20	.035	1.416		5/16 or .3125	12.58
19	.042	1.695	0	.340	13.715
18	.049	1.975		11-32 or .3437	13.875
17	.058	2.35		3/8 or .375	15.10
16	.065	2.637	00	.380	15.26
	1/16 or .0625	2.518		13-32 or .4062	16.34
15	.072	2.92	000	.425	17.125
14	.083	3.35		7/16 or .4375	17.65
	3-32 or .0937	3.78	0000	.454	18.30
13	.095	3.85		15-32 or .4687	18.90
12	.109	4.4	00000	1/2 or .50	20.20

For STEEL PLATES multiply tabular number above (for size), by 1.01.

BUFFALO SCALE CO.

BUFFALO, N. Y.

MANUFACTURERS OF

IMPROVED RAILROAD SCALES

AND SCALES OF ALL DESCRIPTIONS.

Railroad Track, Depot, Platform, Iron, Rolling Mill AND Furnace Charging

Dormant, Hopper, Wheelbarrow, Grain, Hay, Coal AND Cattle

SCALES,

Etc., Etc.

Send for Illustrated Catalogue and Price List. *Every Scale Warranted.* We shall be pleased to correspond with Railroad Managers in relation to prices and terms. ADDRESS

JOHN R. LINEN, SECY., Buffalo, N. Y.

HOWARD IRON WORKS,

MANUFACTURERS OF

BUFFALO POWER HOISTING MACHINES,

HOWARD'S PARALLEL BENCH VISE,

SCHLENKER'S BOLT CUTTER.

Address

R. L. HOWARD & SON,

BUFFALO, N. Y.

Bolt Cutters.

Isaac H. Shearman....Philadelphia, Pa....See Page 80
Phelps & Sanger....Cleveland, Ohio....“ 120
R. L. Howard & Son....Buffalo, N. Y....“ 40
Wm. B. Bement & Son....Philadelphia, Pa....“ 168

Bolt Ends.

Pittsburgh Bolt Co....Pittsburgh, Pa....See Page 118
Plumb, Burdict & Barnard....Buffalo, N. Y....“ 176
Wm. Gilmor of Wm....Baltimore, Md....“ 248

Bolt Heading Machines.

Isaac H. Shearman....Philadelphia, Pa....See Page 80

Bolts.

Collier & Scranton, Oxford Iron Co....New York....See Page 122
G. B. Walbridge....“ “ ....“ 130
Greene & Randolph....“ “ ....“ 54
Hoopes & Townsend....Philadelphia, Pa....“ 28
Jones & Laughlins....Pittsburgh, Pa....“ 32
Lewis Oliver & Phillips; H. B. Newhall, Ag't. .New York....“ 166
Lock Nut & Bolt Co. of New York....“ “ ....“ 172
O. W. Child....“ “ ....“ 262
Phillipsburg Mfg. Co....“ “ ....“ 22
Pittsburgh Bolt Co....Pittsburgh, Pa....“ 118
Plumb, Burdict & Barnard....Buffalo, N. Y....“ 176
Reading Nut & Bolt W'ks; H. B. Newhall, Agt..New York....“ 166
Skinner & Gifford Mfg. Co....Dunkirk, N. Y....“ 140
T. B. Bickerton & Co....Philadelphia, Pa....“ 138
Vose, Dinsmore & Co....New York....“ 60
Wm. Gilmor of Wm....Baltimore, Md....“ 248
Wm. Green & Co....Wilmington, Del....“ 248

Bolt Screwing Machines.

Wm. Sellers & Co....Philadelphia, Pa....See Page 64

Boring Bars.

Wm. Sellers & Co....Philadelphia, Pa....See Page 64

Boring Implements.

Rand & Waring Drill & Compressor Co....New York....See First Page
Wm. P. Kellogg & Co....Troy, N. Y....See Page 98

WEIGHT OF SHEET AND PLATE IRON.—Continued.

Thickness in Inches—Weight of a Square Foot in Pounds.

Inches. Thick.	Lbs. per Square Foot.	Inches Thick.	Lbs per Square Foot.	Inches Thick.	Lbs. per Square Foot.
9/16	22.5	1¾	70.62	3⅞	156.51
5/8	25.21	13/16	73.14	4	161.55
11/16	27.75	7/8	75.58	1/8	166.6
3/4	30.25	15/16	78.20	1/4	171.66
13/16	32.75	2	80.75	3/8	176.71
7/8	35.26	1/8	85.75	1/2	181.77
15/16	37.75	1/4	90.81	5/8	186.79
1	40.35	3/8	95.86	3/4	191.84
1/16	42.87	1/2	100.9	7/8	196.9
1/8	45.4	5/8	105.95	5	201.85
3/16	47.9	3/4	111.	1/8	206.9
1/4	50.45	7/8	116.1	1/4	211.95
5/16	52.96	3	121.15	3/8	217.
3/8	55.45	1/8	126.21	1/2	222.05
7/16	58.01	1/4	131.26	5/8	227.1
1/2	60.52	3/8	136.32	3/4	232.15
9/16	63.05	1/2	141.37	7/8	237.2
5/8	65·56	5/8	146.41	6	242.25
11/16	68.11	3/4	151.46		

For STEEL PLATES multiply tabular number above (for size), by 1.01.

CLEVELAND FOUNDRY
BOWLERS MAHER & BRAYTON
N. P. BOWLER
WM. BOWLER
THOS. MAHER
C. A. BRAYTON
9 11 & 13
WINTER
STREET.

SQUARE BAR IRON.

Weight of a Running Foot, in Pounds.

Thick. Inch's	Wt. per foot. lbs.	Thick. Inch's	Wt. per foot. lbs.	Thick. Inch's	Wt. per foot. lbs.	Thick. Inch's	Wt. per foot. lbs.
1/16	.0131	1 1/16	3.80	2 1/8	15.15	4 1/8	57.20
1/8	.0525	1/8	4.25	1/4	17.	1/4	60.75
3/16	.1182	3/16	4.73	3/8	18.5	3/8	4.55
1/4	.2103	1/4	5.25	1/2	20.5	1/2	68.
5/16	.327	5/16	5.78	5/8	23.1	5/8	72.
3/8	.4735	3/8	6.35	3/4	25.2	3/4	75.65
7/16	.6445	7/16	6.95	7/8	27.5	7/8	79.80
1/2	.84	1/2	7.55	3	30.05	5	83.8
9/16	1.063	9/16	8.2	1/8	32.75	1/8	88.25
5/8	1.314	5/8	8.85	1/4	35.5	1/4	92.5
11/16	1.59	11/16	9.57	3/8	38.25	3/8	97.15
3/4	1.891	3/4	10.30	1/2	41.15	1/2	1 1.
13/16	2.221	13/16	11.05	5/8	44.15	5/8	105.8
7/8	2.575	7/8	11.83	3/4	47.20	3/4	110.5
15/16	2.95	15/16	12.62	7/8	50.25	7/8	115.15
1	3.35	2	13.4	4	53.75	6	120.25

For STEEL multiply tabular number above (for size), by 1.01.

Brass Founders & Finishers.

A. Fulton's Son & Co.....Pittsburgh, Pa.....See Page 116
Du Plaine & Reeves.....Philadelphia, Pa..... " 102
G. W. W. Stevens & Co.....New York..... " 8

Brass Padlocks.

Romer & Co.....Newark, N. J.....See Page 204
Ritchie & Son..... " " ..... " 44

Brass Tubing.

Ansonia Brass & Copper Co.....New York.....See Page 122
Crane Bros. Manufacturing Co.....Chicago, Ill..... " 230

Brass Wire.

Ansonia Brass & Copper Co.....New York.....See Page 122

Brick.

Hall & Sons.....Buffalo, N. Y.....See Page 254

Bridge and Roof Bolts.

Lewis Oliver & Phillips ; H. B. Newhall, Ag't New York.....See Page 166
Phillipsburg Manufacturing Co..... " " ..... " 22
See also, Bridge Irons.

Bridge Builders.

Phillipsburg Manufacturing Co.....New York.....See Page 22
Snyder Brothers.....Williamsport, Pa..... " 46
The McNairy & Claflen Manufacturing Co.....Cleveland, Ohio..... " 42

Bridge Castings.

The Atlas Works.....Pittsburgh, Pa.....See Page 17[illegible]

Bridge Irons.

Greene & Randolph.....New York.....See Page 51
Hoopes & Townsend.....Philadelphia, Pa..... " 28
Jones & Laughlins.....Pittsburgh, Pa..... " 32
Phillipsburg Manufacturing Co.....New York..... " 22
Pittsburgh Bolt Co.....Pittsburgh, Pa..... " 118
Reese, Graff & Woods..... " " ..... " 68

Bridges.

Phillipsburg Manufacturing Co.....New York.....See Page 2[illegible]
The McNairy & Claflen Manufacturing Co.....Cleveland Ohio..... " 4[illegible]

Brokers.

Charles W. Matthews.....Philadelphia, Pa.....See Page 3[illegible]

ROUND BAR IRON.

Weight of a Running Foot, in Pounds.

Diam. Inch's	Wt. per foot. lbs.	Diam. Inch's	Wt. per foot. lbs.	Diam. Inch's	Wt. per foot. lbs.	Diam. Inch's	Wt. per foot. lbs.
1/16	.01	1 1/16	2.975	2 1/8	11.9	4 1/8	44.85
1/8	.0411	1/8	3.338	1/4	13.3	1/4	47.54
3/16	.0925	3/16	3.725	3/8	14.75	3/8	50.33
1/4	.1651	1/4	4.12	1/2	16.41	1/2	53.32
5/16	.2573	5/16	4.545	5/8	18.1	5/8	56.34
3/8	.371	3/8	5.	3/4	19.85	3/4	59.44
7/16	.505	7/16	5.455	7/8	21.5	7/8	62.62
1/2	.657	1/2	5.945	3	23.7	5	65.88
9/16	.835	9/16	6.445	1/8	25.55	1/8	69.23
5/8	1.031	5/8	6.975	1/4	27.81	1/4	72.65
11/16	1.235	11/16	7.52	3/8	29.85	3/8	76.18
3/4	1.475	3/4	8.05	1/2	32.25	1/2	79.75
13/16	1.74	13/16	8.65	5/8	34.45	5/8	83.45
7/8	2.015	7/8	9.25	3/4	37.1	3/4	87.20
15/16	2.317	15/16	9.9	7/8	39.5	7/8	91.05
1	2.625	2	10.55	4	41.95	6	95.

For **STEEL** multiply tabular number above (for size), by 1.01.

M. A. BUELL,

CLEVELAND, OHIO,

MANUFACTURER AND DEALER IN

TELEGRAPH INSTRUMENTS and SUPPLIES.

Instruments, Learner's Instruments,
Batteries, Electric Alarms,
Chemicals, Apparatus,
Materials, Tools.

HOTEL ANNUNCIATORS.

☞ *SEND FOR A CATALOGUE.*

KELLY'S PATENT TURN TABLES.

Centres for Wooden Turn Tables.

PIVOT and SELF-CLOSING CANAL BRIDGES,

Saw Mill Machinery,
Fittings for Oil Tank Cars,
Steam Engines, Boilers, &c.

MANUFACTURED BY

SNYDER BROTHERS,

FOUNDERS and MACHINISTS

WILLIAMSPORT, - PENNSYLVANIA.

Bronze Castings.

Philadelphia Smelting Co............................Philadelphia, Pa...........See Page 134

Bronze Door Knobs.

Romer & Co..Newark, N. J.................See Page 204

Bronze Door Trimmings.

Romer & Co..Newark, N. J.................See Page 204

Brushes for Tubes.

Vose, Dinsmore & Co.................................New York.....................See Page 60

Builders' Hardware.

D. Brewer & Co..Philadelphia, Pa...........See Page 246
England & Bindley......................................Pittsburgh, Pa................ " 162
Geo. Worthington & Co...............................Cleveland, Ohio............. " 100
Sidney Shepard & Co..................................Buffalo, N. Y.................. " 196

Building Materials.

H. W. Johns...New York.....................See Page 224

Bunting.

De Grauw, Aymar & Co...............................New York.....................See Page 112

Burglar Alarms.

Geo. H. Bliss & Co.....................................Chicaco, Ill....................See Page 164

Butts.

England & Bindley......................................Pittsburgh, Pa..............See Page 162
Geo. Worthington & Co...............................Cleveland, Ohio............. " 100
Hermann Boker & Co...................................New York ....See Pages 152, 154
Pratt & Co...Buffalo, N. Y.................See Page 132
Wm. Green & Co...Wilmington, Del............ " 248

HOOP AND SCROLL IRON.

Number of Feet in a Bundle of 5 Pounds.

Hoop Iron.

Size, Width.	Thick.	Feet in Bundle.
5/8 Inches......	No. 21	815
3/4 "	" 20	630
7/8 "	" 19	450
1 "	" 18	360
1 1/8 "	" 17	278
1 1/4 "	" 16	217
1 1/2 "	" 15	160
1 3/4 "	" 15	139
2 "	" 14	110

Scroll Iron.

Size, Width.	Thick.	Feet in Bundle.
1/2 Inches......	No. 10	240
5/8 "	" 16	430
5/8 "	" 14	347
5/8 "	" 10	190
3/4 "	" 16	360
3/4 "	" 14	290
3/4 "	" 12	208
3/4 "	" 10	160
7/8 "	" 16	310
7/8 "	" 14	249
7/8 "	" 12	175
1 "	" 16	270
1 "	" 14	216
1 "	" 12	152

WEIGHT OF TIRE IRON PER SET OF 54 FEET.

Size.	Pounds.
1 by 3/16	34
1 " 1/4	45
1 " 5/16	56
1 " 3/8	68
1 1/8 " 1/4	50
1 1/8 " 5/16	63
1 1/8 " 3/8	75
1 1/8 " 7/16	88
1 1/8 " 1/2	101
1 1/4 " 1/4	56
1 1/4 " 5/16	70
1 1/4 " 3/8	85
1 1/4 " 7/16	99
1 1/4 " 1/2	113

Size.	Pounds.
1 3/8 by 3/8	93
1 3/8 " 1/2	124
1 1/2 " 3/8	101
1 1/2 " 1/2	135
1 1/2 " 5/8	169
1 5/8 " 1/2	148
1 5/8 " 5/8	183
1 3/4 " 1/2	158
1 3/4 " 5/8	197
1 3/4 " 3/4	236
2 " 1/2	180
2 " 5/8	225
2 " 3/4	270

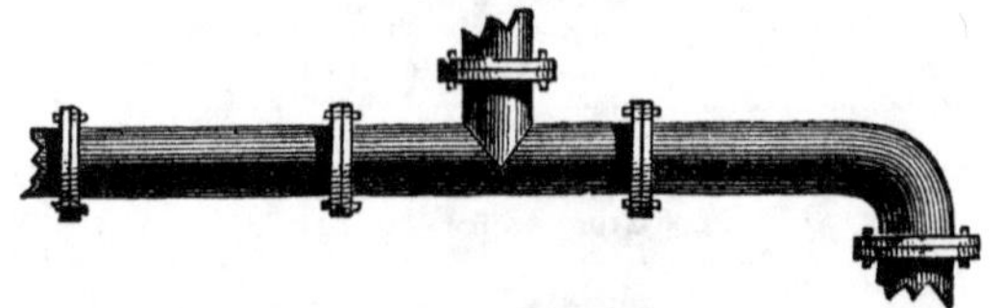

CAST IRON FLANGE PIPES

BOTH FACED AND DRILLED AND PLAIN.

GAS and WATER PIPES,

AND CONNECTIONS FOR SAME.

Valves, Fire Hydrants, &c.

R. A. BRICK & CO.

NO. 112 LEONARD ST., NEW YORK.

JAMES E. GRANNISS. JOHN MERRY.

JOHN MERRY & CO.

West Side Galvanizing Works,

541, 543, 545 & 547 West 15th St.,

Office and Warehouse, No. 46 Cliff St., New York.

MANUFACTURERS OF

BEST BLOOM AND REFINED

AMERICAN GALVANIZED SHEET IRON,

CORRUGATED IRON FOR ROOFING,

GALVANIZED TELEGRAPH AND FENCE WIRE,

GALVANIZED AND TINNED NAILS,

ALL KINDS OF IRON WORK GALVANIZED OR TINNED TO ORDER.

Buyers of Old Metal.

Du Plaine & Reeves.....Philadelphia, Pa.....See Page 102
Hooks' Smelting Co..... " " ..... " 216

Cabinet Makers' Machinery.

C. B. Rogers & Co.....New York.....See Page 210
J. A. Fay & Co.....Cincinnati, Ohio..... " 206

Cabinet Woods.

Joseph Churchyard.....Buffalo, N. Y.....See Page 160

Cable Chain.

De Grauw, Aymar & Co.....New York.....See Page 112
Hermann Boker & Co..... " " .....See Pages 152, 154

Cables—Telegraph.

Bishop Gutta Percha Works.....New York.....See Page 34
Charles T. Chester..... " " ..... " 212
Leclanche Battery Co..... " " ..... " 96

Callipers.

Wm. Sellers & Co.....Philadelphia, Pa.....See P ge 64

Canal Bridges—Self-Closing.

Snyder Brothers.....Williamsport, Pa.....See Page 46

CAST IRON FLANGE PIPES.

(*Furnished by R. A. Brick & Co.*)

Internal Diameter. Feet.	Usual Thickness. Inches.	Weight per foot. Pounds.	Length. Feet.	Usual Diameter: Flange-Inches.
2	$\frac{3}{8}$ scant	8 to 10	$8\frac{1}{2}$ or less	6
3	$\frac{3}{8}$ full	13 to 16	"	$7\frac{1}{2}$
4	$\frac{1}{2}$ scant	20 to 24	"	$8\frac{1}{2}$
5	$\frac{1}{2}$ full	25 to 30	"	$9\frac{1}{2}$
6	$\frac{5}{8}$ scant	33 to 39	"	11
8	$\frac{5}{8}$ full	48 to 55	"	13
10	$\frac{3}{4}$ scant	65 to 75	"	15
12	$\frac{3}{4}$ full	90 to 100	"	18

CAST IRON COLUMNS.

Weight that can be borne with Safety by Cast-Iron Columns in* 1000 *Lbs.

(Trenton Iron Works.)

Diam. in Ins.	Length of Column in Feet.										
	5 ft.	6 ft.	7 ft.	8 ft.	9 ft.	10 ft.	12 ft.	14 ft.	16 ft.	18 ft.	20 ft.
2	12.4	9.4	7.2	—	—	—	—	—	—	—	—
3	44	36	30	24	20	18	—	—	—	—	—
4	102	88	76	66	56	48	38	28	—	—	—
5	184	164	146	130	114	102	80	64	52	44	—
6	288	264	242	218	198	180	136	122	100	84	72
7	414	386	360	332	306	282	238	200	170	144	124
8	560	532	502	470	440	410	354	304	262	226	196
9	728	698	660	630	596	560	494	432	378	332	292
10	916	884	850	812	774	739	658	586	520	462	410
11	1126	1082	1056	1016	974	932	846	774	686	616	552
12	1354	1320	1281	1240	1196	1152	1056	966	878	796	720
13	—	1570	1530	1486	1440	1392	1292	1192	1094	1000	912
14	—	—	1798	1754	1706	1656	1550	1440	1332	1228	1130
15	—	—	2086	2040	1992	1940	1828	1712	1596	1482	1372

The above table is based upon the following conditions:—The column must be placed precisely perpendicular; both ends must be faced exactly at a right angle with the vertical axis, and the load must be evenly distributed over the whole face.

If both ends are rounded, the weight should be reduced to one-third of the above; if one end is rounded, two-thirds should be allowed.

Candles.

F. S. Pease........................Buffalo, N. Y..................See Page 100
L. G. Tillotson & Co........................New York.................... " 94
P. H. Moffat........................Chicago, Ill.................... " 190

Cap Screws.

Worcester Machine Screw Co.; H. B. Newhall, Agent........................New York......................See Page 166

Car Axles.

De Laney & Co........................Buffalo. N. Y.................See Page 98
Jonas S. Heartt & Co........................Troy, N. Y.................... " 134
Pittsburgh Forge & Iron Co........................Pittsburgh, Pa............... " 174

WIRE—IRON, STEEL, COPPER, BRASS.

Weight of 100 Feet in Pounds.

(Taken from Haswell.)

American Wire Gauge.					Birmingham Wire Gauge.				
No. of Gauge.	Wire—per Lineal Foot.				No. of Gauge.	Wire—per Lineal Foot.			
	Iron.	Steel.	Copper.	Brass.		Iron.	Steel.	Copper.	Brass.
0000	56.07	56.60	64.05	60 52	0000	54.62	55.13	62.39	58.93
000	44 47	44.89	50.79	47.99	000	47.86	48.32	54.67	51.64
00	35.26	35 6	40.28	38.07	00	38.27	38.63	43.71	41.28
0	27.97	28.23	31.94	30.18	0	30.63	30.92	34.99	33.05
1	22.18	22.39	25.33	23.93	1	23.85	24.07	27.24	25.73
2	17.59	17.75	20 09	18.98	2	21.37	21.57	24.41	23.06
3	13.95	14.08	15.93	15.05	3	17 78	17.94	20.3	19.18
4	11.06	11.17	12 63	11.94	4	15.01	15.15	17.15	16.19
5	8.772	8.855	10.02	9 467	5	12.82	12 95	14.65	13.84
6	6.957	7.022	7.946	7.507	6	10.92	11.02	12.47	11.78
7	5.552	5. 68	6.301	5.954	7	8.586	8.667	9.807	9.263
8	4.375	4.416	4. 98	4.722	8	7.214	7.283	8.241	7.783
9	3.47	3.503	3.964	3.744	9	5.805	5.859	6.63	6.262
10	2.751	2.777	3. 43	2 969	10	4.758	4.803	5.435	5.133
11	2.182	2.203	2.492	2.355	11	3.816	3.852	4.359	4 117
12	1.730	1 747	1.977	1.867	12	3.148	3.178	3.596	3.397
13	1.372	1.385	1.567	1.481	13	2. 92	2.414	2.732	2.58
14	1.089	1.099	1.213	1.174	14	1. 26	1 843	2.085	1.969
15	.8631	.8712	.9859	.9315	15	1.374	1.387	1.569	1.482
16	.6845	.6909	.7819	. 587	16	1.119	1.13	1.279	1.208
17	.5427	.5478	.6 99	.5857	17	.8915	.9	1.018	.9618
1	.4304	.4344	.4916	.4645	18	.6363	.6423	.7268	.6864
19	.3413	.3445	.3899	.3683	19	.4675	.472	.534	.5043
20	.2708	.2734	.3094	.292	20	.3246	.3277	.3709	.3502
21	.2147	.2167	.2452	.2317	21	.2714	.274	.31	.2929
22	.1703	.1719	.1945	.1838	22	.2079	.2098	.2373	.2241
23	.135	.1363	.1542	.1457	23	.1656	.1672	.1892	.1788
24	.1071	.1081	.1223	.1155	24	.1283	.1295	.1465	.1384
25	.0849	.0857	.097	.0916	25	.106	.107	.1211	.1144
26	.0673	.068	.0769	.0727	26	.0859	.0867	.0981	.0926
27	.0534	.0539	.061	.0576	27	.0678	.0685	.0775	.0732
28	.0423	.0427	.0484	.0457	28	.05 9	.0524	.0593	.056
29	.0336	.0339	.0383	.036	29	.0448	.0452	.0511	.0483
30	.02 6	.02 9	.0304	.0287	30	.0382	.0385	.0436	.0412
31	.0211	.0213	.0241	.0228	31	0265	.0267	.0303	.0286
32	.01 7	.0169	.0 91	.0181	32	.0215	.0217	.0245	.0231
33	.01 3	.0134	.0152	.0143	33	.017	.0171	.0194	.0183
34	.0105	.0106	.0120	.0114	34	.013	.0131	.0148	.014
35	.00 3	.0084	.0 96	. 09	35	.0066	.0067	.0076	.0071
36	.0066	.0067	.0076	.0071	36	.0042	.0043	.0048	.0046

ESTABLISHED 1782.

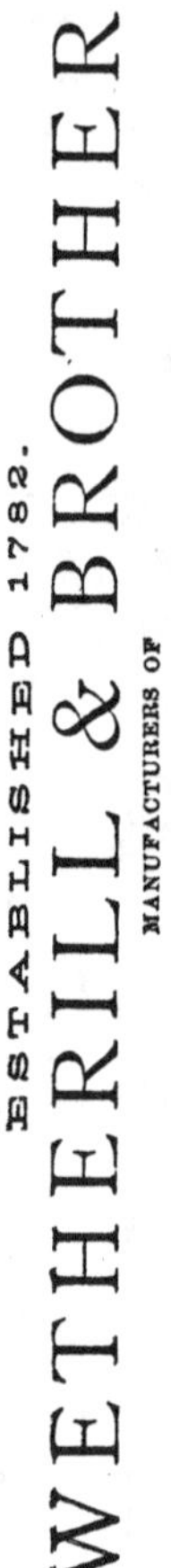

MANUFACTURERS OF

WHITE LEAD, RED LEAD, LITHARGE AND ORANGE MINERAL.

The present proprietors succeed Wm. Wetherill, successor to John P. & Wm. Wetherill, sons of Sam'l Wetherill, the Original Manufacturer of the above articles in the United States.

Office, No. 41 North Front St. **PHILADELPHIA.** *Factory, 30th St. below Chestnut.*

Car and Journal Bearings.

A. Fulton's Son & Co..........Pittsburgh, Pa..........See Page 116
Du Plaine & Reeves..........Philadelphia, Pa.........." 102
Hooks Smelting Co.......... " " ...." 216
G. W. W. Stevens & Co..........New York.........." 8

Car Bits.

Champlin & Rogers..........Chicago, Ill..........See Page 192

Car Boxes.

Isaac U. Coles..........New York..........See Page 20

Car Box Borers.

Buffalo Machinery Agency..........Buffalo, N. Y..........See Page 160
Vose, Dinsmore & Co..........New York.........." 60

Car Brasses.

Philadelphia Smelting Co..........Philadelphia, Pa..........See Page 134

Car Builders.

Erie Car Works..........Erie, Pa..........See Page 188
The McNairy & Claflen Manufacturing Co..........Cleveland, Ohio.........." 42
Wason Manufacturing Co..........Springfield, Mass.........." 146

Car Colors.

C. E. Hecht..........Easton, Pa..........See Page 158
Edward Smith & Co..........New York.........." 218
Jno. W. Masury & Son.......... " " ...." 214

ACRES REQUIRED FOR DIFFERENT WIDTHS OF ROADWAY.

(Condensed from Trautwine.)

Width of Road in Feet.	Acres per Mile.	Acres per 100 Ft.	Width of Road in Feet.	Acres per Mile.	Acres per 100 Ft.
1	.121	.002	48	5.82	.11
2	.242	.005	49½	6.	.114
3	.364	.007	50	6.06	.115
4	.485	.009	52	6.3	.119
5	.606	.011	54	6.55	.124
6	.727	.014	56	6.79	.129
7	.848	.016	57¾	7.	.133
8	.970	.018	58	7.03	.133
¼	1.	.019	60	7.27	.138
9	1.09	.021	62	7.52	.142
10	1.21	.023	64	7.76	.147
12	1.46	.028	66	8.	.151
14	1.70	.032	68	8.24	.156
16	1.94	.037	70	8.48	.161
½	2.	.038	72	8.73	.165
18	2.18	.041	74	8.97	.17
20	2.42	.046	¼	9.	.17
22	2.67	.051	76	9.21	.174
24	2.91	.055	78	9.45	.179
¾	3.	.057	80	9.7	.184
26	3.15	.06	82	9.94	.188
28	3.39	.064	½	10.	.189
30	3.64	.069	84	10.2	.193
32	3.88	.073	86	10.4	.197
34	4.12	.078	88	10.7	.202
36	4.36	.083	90	10.9	.207
38	4.61	.087	¾	11.	.209
40	4.85	.092	92	11.2	.211
41¼	5.	.094	94	11.4	.216
42	5.09	.096	96	11.6	.22
44	5.33	.101	98	11.9	.225
46	5.58	.106	100	12.1	.23

Car Door Catches.

Posts & Kalkman...New York........................See Page 266

Car Door Locks.

Romer & Co...Newark, N. J..................See Page 204

Car Equipments.

Du Plaine & Reeves.......................................Philadelphia, Pa.............See Page 1[illegible]2

Car Glass.

D. R. Hobart & Co..New York........................See Page 126
Leffingwell & Co ....Cleveland, Ohio............... " 190
L. G. Tillotson & Co. ....New York....................... " 94

HOLLOW CAST IRON COLUMNS (CYLINDRICAL).

Breaking Weight in Tons.

This is a *comparative* table of Breaking Weights; the columns headed **H.** as given by Hurst, and the columns headed **T.** as given by Trautwine. Trautwine's figures are qualified by many explanations and remarks on the different qualities of iron and the various tests and experiments on which the rules, by which the following figures were made, are based. The table will serve to show the great difference of opinion, and results, which skilled men and minds will entertain on the same subject.

Ext'rn'l Diam.	Thick. of Metal	Length of Column in Feet.									
		8 Feet.		10 Feet.		12 Feet.		14 Feet.		16 Feet.	
		H.	**T.**	**H.**	**T.**	**H.**	**T.**	**H.**	**T.**	**H.**	**T.**
3	½	40	50	32	—	23	—	18	—	14	—
3	⅝	47	—	35	—	26	—	20	—	16	—
3	¾	54	—	38	—	28	—	22	—	17	—
3½	½	59	72	51	55	36	—	27	—	23	—
3½	⅝	71	—	53	—	42	—	32	—	25	—
3½	¾	81	92	62	68	44	—	35	—	26	—
4	½	81	98	61	76	47	59	36	—	34	—
4	⅝	92	—	73	—	56	—	44	—	39	—
4	¾	113	129	85	100	65	76	48	—	38	—
4	1	139	—	104	—	80	—	64	—	48	—
4½	½	106	127	81	101	65	81	50	—	44	—
4½	⅝	128	—	99	—	77	—	61	—	55	—
4½	¾	149	169	115	134	89	108	72	—	60	—
4½	1	185	—	143	—	111	—	88	—	71	—
5	½	133	157	104	126	83	104	67	—	54	—
5	⅝	161	—	127	—	91	—	81	—	70	—
5	¾	188	214	148	171	117	140	90	—	77	—
5	1	236	—	186	—	148	—	119	—	96	—
5½	½	153	190	129	155	105	129	85	108	70	—
5½	⅝	187	—	157	—	128	—	104	—	88	—
5½	¾	218	261	184	212	149	175	121	147	102	—
5½	1	276	—	232	—	189	—	153	—	127	—
6	½	190	224	155	185	127	155	95	131	87	—
6	⅝	232	—	190	—	156	—	128	—	106	—
6	¾	272	311	223	256	183	213	150	180	125	—
6	1	345	—	283	—	232	—	191	—	159	—
6½	⅝	269	—	224	—	187	—	152	—	130	—
6½	¾	316	363	263	303	219	255	178	217	153	—
6½	1	403	—	336	—	280	—	228	—	196	—
7	⅝	307	—	260	—	219	—	185	—	156	—
7	¾	361	416	306	351	258	298	217	255	184	221
7	1	462	522	391	438	330	370	278	316	236	272
8	1	583	667	507	571	438	491	377	425	325	370
9	1	705	818	627	713	553	616	481	545	423	480

IMPROVED AIR BRAKE,

MANUFACTURED BY THE

Ward Air Brake Co.

OF KALAMAZOO, MICHIGAN.

CHARTERED BY THE STATE OF MICHIGAN, FEBRUARY, 1872.

A trial of the *Ward Air Brake* for more than one year has fully demonstrated it to be the CHEAPEST, SIMPLEST and most effective Air Brake now in use. It has much fewer parts, a more speedy release, and costs nearly one-half less than other Air Brakes. Has been in constant use for over one year, and is most highly recommended by prominent railroad men.

A trial train equipped upon any road, to be paid for only when found satisfactory. ☞ Full particulars and terms furnished upon application.

L. B. KENDALL, *President.* **T. S. COBB;** *Sec. and Treas.*

LUCIUS HART & Co.

Importers & Dealers in

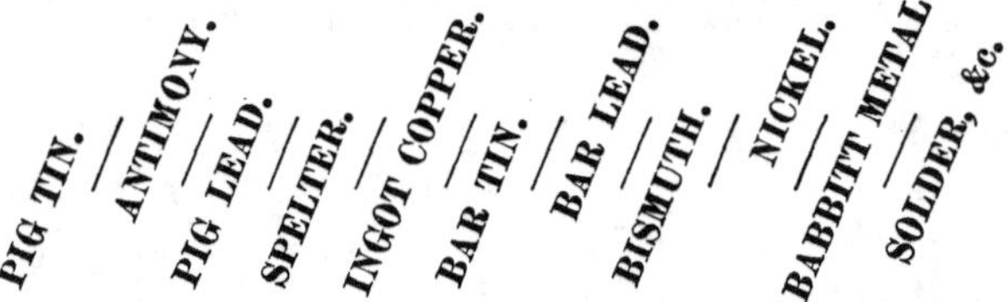

8 & 10 BURLING SLIP,

CHAS. FANNING.
LUCIUS HART.

NEW YORK.

J. D. WEST & CO.

40 Cortlandt Street, - New York.

WEST'S IMPROVED AND OTHER

PUMPS,

WITH PIPE, HOSE, &c.

WEST'S COPPER, OTIS' AND OTHER

Lightning Rods,

WEATHER VANES, CRESTINGS, &c.

Car Hassocks.

Geo. Drake Smith..Philadelphia, Pa.............See Page 246
New York Ottoman & Hassock Co..................New York....................... " 236

Car Hooks.

Vose, Dinsmore & Co...................................New York.......................See Page 60

Car Irons.

Greene & Randolph......................................New York.......................See Page 54
Hoopes & Townsend.....................................Philadelphia, Pa............. " 28

Car Linings.

E S. Lunt...New York.......................See Page 84
L. G. Tillotson & Co..................................... " " .... " 94

Car Makers' Machinery.

C. B. Rogers & Co..New York.......................See Page 210
J. A. Fay & Co..Cincinnati, Ohio.............. " 2[illegible]6
Richards, London & Kelley...........................Philadelphia, Pa............. " 66

Car Packing.

Isaac U. Coles..New York.......................See Page 20

HEAVY WROUGHT IRON ARTESIAN TUBES.

Inside Diameter.	Weight per foot, lbs.	Inside Diameter.	Weight per foot, lbs.
1½	2.694	4½	12.492
2	3.667	5	14.564
2½	5.773	6	18.767
3	7.547	7	23.410
3½	9.055	8	28.348
4	10.728		

CAPACITY OF CISTERN, IN GALLONS, FOR EACH FOOT IN DEPTH.

Diam. in feet.	Gallons.	Diam. in feet.	Gallons.	Diam. in feet.	Gallons.	Diam. in feet.	Gallons.
2.	23.4	5.5	177.7	9.	475.87	15.	1,321.9
2.5	36.6	6.	211.5	9.5	553.67	20.	2,350.1
3.	43.0	6.5	248.22	10.	587.5	25.	3,570.7
3.5	71.50	7.	287.79	11.	710.9	30.	5,287.7
4.	94.30	7.5	330.48	12.	845.	35.	7,189.
4.5	119.	8.	376.	13.	992.9	40.	9.367.2
5.	147.	8.5	424.44	14.	1151.5	45.	11,893.2

CEMENT TO RESIST FIRE AND WATER, AND HARDEN QUICKLY.

Two parts finely sifted unoxodized iron filiugs.
Ono part, perfectly dry, finely powdered loam.
Knead the mixture with strong vinegar into a homogeneous plastic mass, to be used as soon as made.

CEMENT FOR STOPPING JOINTS, ETC.

White lead in oil, mixed with enough white sand to make a stiff paste. This grows hard by exposure, aud resists heat, cold and water.

Car Plushes, &c.

E. S. Lunt..........New York..........See Page 84
L. G. Tillotson & Co.......... " " .... " 94
Vose, Dinsmore & Co.......... " " .... " 60

Carpenters' Tools.

England & Bindley..........Pittsburgh, Pa..........See Page 162
Geo. Worthington & Co..........Cleveland, Ohio.......... " 100
Hermann Boker & Co..........New York..........See Pages 152, 154

Carriage Builders' Machinery.

C. B. Rogers & Co..........New York..........See Page 210

Carriage Bolts.

Lewis Oliver & Phillips; H. B. Newhall, Agt...New York..........See Page 166

Carriage Hardware, Wheels, Spokes, &c.

Wm. Green & Co..........Wilmington, Del..........See Page 248

WEIGHTS AND DIMENSIONS OF GAS PIPES.

Inside Diameter in inches.	Outside Diameter in inches.	Weight per foot in pounds.	Inside Diameter in inches.	Outside Diameter in inches.	Weight per foot. in pounds.
1/8	0.40	0.24	3	3.5	7.54
1/4	0.[illegible]4	0.42	3½	4.0	9.05
3/8	0.67	0.56	4	4.5	10.72
1/2	0.84	0.85	4½	5.0	12.49
3/4	1.05	1.12	5	5.56	14.56
1	1.31	1.67	6	6.62	18.77
1¼	1.66	2.25	7	7.62	23.41
1½	1.95	2.69	8	8.62	28.35
2	2.[illegible]7	3.66	9	9.68	34.07
2½	2.87	5.77	10	10.75	40.64

WEIGHTS AND DIMENSIONS OF LAP-WELDED IRON BOILER FLUES.

Outside Diameter	Thickness W. G.	Lbs. weight per foot.	Outside Diameter	Thickness W. G.	Lbs. weight per foot.
1¼	14	1.65	3¼	11	4.15
1½	14	1.70	3½	10	5.20
1¾	13	1.85	3¾	10	5.30
2	13	2.10	4	10	5.55
2¼	13	2.30	5	9	7.1
2½	12	2.50	6	8	10.5
2¾	12	3.15	7	8	12.2
3	11	3.60			

LIGHT WROUGHT IRON ARTESIAN TUBE AND CASING FOR OIL WELLS

Standard Sizes.

Outside Diameter. Inches.	Inside Diameter. Inches.	Weight per Foot. Pounds.	Outside Diameter. Inches.	Inside Diameter. Inches.	Weight, per Foot. Pounds.
1¾	1½	1.665	4¼	4	5.500
2¼	2	2.238	4½	4¼	6.[illegible]10
2½	2¼	2.755	5	4¾	7.226
2¾	2½	3.045	5¼	5	7.667
3	2¾	3.333	5½	5 3/16	8.083
3¼	3	3.958	6	5⅝	9.346
3½	3¼	4.272	6⅝	6¼	1[illegible].064
3¾	3½	4.950	7	6⅝	12.435
4	3¾	5.320	8	7⅝	15.[illegible]09
			8⅝	8¼	16.155

Cars.

Chas. W. Matthews	Philadelphia, Pa	See Page 38
Erie Car Works	Erie, Pa	" 188
Miller & Smith	New York	" 112
O. W. Child	" "	" 262
Skinner & Gifford Mfg. Co	Dunkirk, N. Y	" 140
The McNairy & Claflen Mfg. Co	Cleveland, Ohio	" 42
Wason Mfg. Co	Springfield, Mass	" 146

See also, Car Builders.

Car Springs.

Chicago Spring Works	Chicago, Ill	See Page 200
Goodyears I. R. G. Mfg. Co	New York	" 252
New York Rubber Co	" "	" 260
Posts & Kalkman	" "	" 266
Vose, Dinsmore & Co	" "	" 60

Car Stoves.

L. G. Tillotson & Co	New York	See Page 94

Car Trimmings.

L. G. Tillotson & Co	New York	See Page 94

Car Varnishes.

Edward Smith & Co	New York	See Page 218
Valentine & Co	" "	" 198

Car Wheel Borers.

Wm. Sellers & Co	Philadelphia, Pa	See Page 64

Car Wheels.

See Wheels.

RUBBER CAR SPRINGS,

To resist actual compression and bulging, should be used in layers, with inter vening division plates; if used in bulk, should have retaining rings to preven bursting.

The duration of spring depends, of course, on its quality. Pure rubber is of little use; it must be vulcanized. It is adulterated with plumbago, though in best springs as much as 40 per cent. foreign substances are used—principally sulphide of lead and soapstone.

WEIGHT OF RUBBER CAR SPRINGS.

Diameter in Inches.	Height in Inches.	Diameter of Hole.	Weight.
4	4	1	3 lbs.
4½	4½	1	4 lbs. 5 oz.
5	5	1	5½ "
5½	5½	1	7 " 10 oz.
6	6	1	9¾ "
6½	6½	1	12 " 5 oz.
7	7	1	15¾ "
7½	7½	1	19½ "
8	8	1	23½ "
8½	8½	1	27¾ "
9	9	1	32¼ "
10	10	1	45½ "
11	11	1	59 "

Castings.

Bowlers, Maher & Brayton.............................Cleveland, Ohio.............See Page 42
Skinner & Gifford Mfg. Co.............................Dunkirk, N. Y................ " 140
The Atlas Works..Pittsburgh, Pa............... " 172

Cast Iron Sinks.

R. L. Howard & Son......................................Buffalo, N. Y...................See Page 40

SHEET AND BAR BRASS.

Weight in Pounds.

Thickness, or Diameter, or Side; Inches.	Sheets per Square Foot.	Square Bars 1 Foot Long.	Round Bars 1 Foot Long.	Thickness, or Diameter, or Side; Inches.	Sheets per Square Foot.	Square Bars 1 Foot Long.	Round Bars 1 Foot Long.
1/16	2.7	.015	.011	1 1/16	45.95	4.08	3.20
1/8	5.41	.055	.045	1/8	48.69	4.55	3.57
3/16	8.12	.125	.1	3/16	51.4	5.08	3.97
1/4	10.76	.225	.175	1/4	54.18	5.65	4.41
5/16	13.48	.350	.275	5/16	56.85	6.22	4.86
3/8	16.25	.51	.395	3/8	59.55	6.81	5.35
7/16	19.	.69	.54	7/16	62.25	7.45	5.85
1/2	21.65	.905	.71	1/2	65.	8.13	6.37
9/16	24.3	1.15	.9	9/16	67.75	8.83	6.92
5/8	27.12	1.4	1.1	5/8	70.35	9.55	7.48
11/16	29.77	1.72	1.35	11/16	73.	10.27	8.05
3/4	32.46	2.05	1.6	3/4	75.86	11.	8.65
13/16	35.18	2.4	1.85	13/16	78 55	11.82	9.29
7/8	37.85	2.75	2.15	7/8	81.25	12.68	9 95
15/16	40.55	3.15	2.48	15/16	84.	13.5	10.58
1	43.29	3.65	2.85	2	86.75	14.35	11.25

ALLOYS.

Alloys.	Zinc.	Tin.	Copper.	Antimony.	Lead.	Bismuth
Babbitt meta	......	10	1	1	......	......
Bell metal	......	5	16	......	......	......
Brass engine bearings	1/4	13	112	......	......	......
" (tough) for engine work	15	15	1 0	......	......	......
" " for heavy bearings	5	25	160	......	......	......
" , yellow for turning	1	......	2	......	......	......
" for locomotive bearings	1	7	64	......	......	......
" for straps and glands	1	16	130	......	......	......
Flanges to stand brazing	1	......	32	......	1	......
Metal to expand in cooling	......	......	......	2	9	1
Muntz's sheathing	40	......	60	......	......	......
Pewter	......	100	......	17	......	......
Spelter	1	......	1	......	......	......
Statuary bronze	5	2	90	......	2	......
Solders.						
For lead	......	1	......	......	1½	......
" tin	......	1	......	......	2	......
" pewter	......	2	......	......	1	......
" brazing (soft)	3	1	4	......	......	......
" " "	......	2	......	1	......	......
" " (hard)	1	......	1	......	......	......
" " (hardest)	1	......	3	......	......	......

ESTABLISHED 1848.

WM. SELLERS & CO.

1600 HAMILTON ST., PHILADELHIA.

Engineers and Machinists.

MANUFACTURERS OF

IRON AND STEEL

WORKING

TOOLS

FOR

Railways,

Machine Shops,

Boiler Makers,

Ship Yards, &c.

Our Tools have been constructed with special reference to *economy* and *convenience of working*. Many of them are of novel and patented designs.

PIVOT BRIDGES and RAILWAY TURN TABLES.

MILL GEARING AND SHAFTING A SPECIALTY.

PATENTEES AND MANUFACTURERS OF

INJECTORS FOR FEEDING BOILERS.

☞ Full descriptive specifications, photographs and prices sent on application.

Cast Steel.

Atlantic Steel Works........New York........See Page 212
Atwater, Wheeler & Co........New Haven, Conn........ " 10
Geo. Worthington & Co........Cleveland, Ohio........ " 100
Reese, Graff & Woods........Pittsburgh, Pa........ " 68
T. B. Bickerton & Co........Philadelphia, Pa........ " 138
Tyng & Co........New York........ " 260
Vose, Dinsmore & Co........ " " .... " 60

Cement.

F. O. Norton........New York........See Page 44
H. W. Johns........ " " .... " 224
Van Tuyl Manufacturing Co........ " " .... " 242

Chain Links.

Hoopes & Townsend........Philadelphia, Pa........See Page 28

Chains.

De Grauw, Aymar & Co........New York........See Page 112
Geo. Worthington & Co........Cleveland, Ohio........ " 100
Hermann Boker & Co........New York........See Pages 152, 154
T. B. Bickerton & Co........Philadelphia, Pa........See Page 138
Tyng & Co........New York........ " 260
Vose, Dinsmore & Co........New York........ " 60

Chains, Measuring.

F. Eckel........New York........See Page 76

Chasing Lathes.

Wm. Sellers & Co........Philadelphia, Pa........See Page 64

WEIGHT OF FLAT IRON.—4 PAGES.

Weight of Running Foot in Pounds.

Thickness in Inches.

Width in Ins.	1/16	1/8	3/16	1/4	5/16	3/8
1	.21	.41	.62	.83	1.04	1.25
1/8	.23	.47	.7	.94	1.17	1.41
1/4	.26	.52	.78	1.04	1.3	1.56
3/8	.29	.57	.86	1.14	1.43	1.72
1/2	.31	.62	.94	1.25	1.56	1.87
5/8	.34	.68	1.01	1.35	1.69	2.03
3/4	.36	.73	1.09	1.46	1.82	2.19
7/8	.39	.78	1.17	1.56	1.95	2.34
2	.42	.83	1.25	1.67	2.08	2.5
1/8	.44	.88	1.33	1.77	2.21	2.65
1/4	.47	.94	1.4	1.87	2.34	2.81
3/8	.5	.99	1.48	1.98	2.47	2.97
1/2	.52	1.04	1.56	2.09	2.6	3.12
5/8	.55	1.09	1.64	2.19	2.73	3.28
3/4	.57	1.14	1.72	2.29	2.86	3.44
7/8	.6	1.2	1.8	2.4	2.99	3.59
3	.62	1.25	1.87	2.5	3.12	3.75
1/8	.65	1.3	1.95	2.6	3.26	3.91
1/4	.68	1.35	2.03	2.7	3.38	4.06
3/8	.7	1.4	2.11	2.81	3.52	4.22
1/2	.73	1.46	2.19	2.91	3.65	4.37
5/8	.76	1.51	2.27	3.02	3.78	4.53
3/4	.78	1.56	2.34	3.12	3.91	4.69
7/8	.81	1.61	2.42	3.23	4.03	4.84
4	.83	1.66	2.5	3.33	4.17	5.00
1/8	.86	1.72	2.58	3.44	4.3	5.16
1/4	.88	1.77	2.66	3.54	4.43	5.31
3/8	.91	1.82	2.73	3.64	4.56	5.47
1/2	.94	1.87	2.81	3.75	4.69	5.62
5/8	.96	1.93	2.89	3.85	4.82	5.78
3/4	.99	1.98	2.97	3.96	4.95	5.94
7/8	1.01	2.03	3.05	4.06	5.08	6.1
5	1.04	2.08	1.12	4.17	5.21	6.25
1/8	1.06	2.13	3.2	4.27	5.34	6.41
1/4	1.1	2.19	3.28	4.37	5.47	6.56

Chemicals—Telegraph.

Leclanche Battery Co..........New York..........See Page 96
See also, Telegraph Materials.

Chilled Faced R. R. Frogs.

Bowlers, Maher & Brayton..........Cleveland, Ohio..........See Page 42

Chimes.

A. Fulton's Son & Co..........Pittsburgh, Pa..........See Page 116

Chisels.

Hermann B ker & Co..........New York..........See Pages 152, 154
See also Carpenters' and Mechanics' Tools.

Chucks.

Champlin & Rogers..........Chicago, Ill..........See Page 192
D. Brewer & Co..........Philadelphia, Pa.........." 246

Church and School Bells.

A. Fulton's Son & Co..........Pittsburgh, Pa..........See Page 116

Chute Nails.

D. Brewer & Co..........Philadelphia, Pa..........See Page 246

Circular Saw Mills.

Erie City Iron Works..........Erie, Pa..........See Page 170

Circular Saws.

American Saw Co..........New York..........See Page 156
Buffalo Machinery Agency..........Buffalo, N. Y.........." 160
Erie City Iron Works..........Erie, Pa.........." 170

WEIGHT OF FLAT IRON.—CONTINUED.

Weight of Running Foot in Pounds.

Width in Ins.	Thickness in Inches.					
	1/16	1/8	3/16	1/4	5/16	3/8
5 3/8	1.12	2.24	3.36	4.48	5.6	6.72
1/2	1.14	2.29	3.44	4.58	5.73	6.88
5/8	1.17	2.34	3.52	4.69	5.86	7.03
3/4	1.2	2.39	3.59	4.79	6.	7.19
7/8	1.22	2.45	3.67	4.9	6.12	7.34
6	1.25	2.5	3.75	5.	6.25	7.5
1/8	1.27	2.55	3.83	5.1	6.39	7.66
1/4	1.3	2.6	3.91	5.2	6.51	7.81
3/8	1.32	2.66	3.98	5.31	6.64	7.97
1/2	1.35	2.7	4.06	5.41	6.77	8.13
5/8	1.38	2.76	4.14	5.52	6.91	8.29
3/4	1.4	2.81	4.22	5.62	7.03	8.44
7/8	1.43	2.86	4.3	5.73	7.16	8.59
7	1.46	2.92	4.37	5.83	7.29	8.75
1/4	1.51	3.02	4.53	6.04	7.55	9.07
1/2	1.56	3.12	4.69	6.25	7.81	9.37
3/4	1.61	3.23	4.84	6.46	8.07	9.69
8	1.67	3.33	5.	6.67	8.33	10.
1/4	1.72	3.43	5.16	6.87	8.6	10.3
1/2	1.77	3.54	5.32	7.08	8.85	10.63
3/4	1.82	3.65	5.47	7.29	9.11	10.94
9	1.87	3.75	5.62	7.5	9.37	11.25
1/4	1.93	3.85	5.78	7.71	9.63	11.56
1/2	1.98	3.96	5.94	7.92	9.89	11.87
3/4	2.03	4.06	6.09	8.12	10.15	12.19
10	2.08	4.17	6.25	8.33	10.4	12.5
1/4	2.13	4.27	6.4	8.54	10.67	12.8
1/2	2.19	4.37	6.56	8.75	10.93	13.13
3/4	2.24	4.48	6.72	8.96	11.20	13.43
11	2.29	4.58	6.87	9.16	11.45	13.75
1/4	2.34	4.69	7.03	9.37	11.72	14.06
1/2	2.39	4.79	7.18	9.58	11.97	14.37
3/4	2.45	4.89	7.34	9.79	12.25	14.69
12	2.5	5.	7.5	10.	12.5	15.

REESE, GRAFF & WOODS,

MANUFACTURERS OF

Bar Iron,
Refined Iron,
Sheet Iron,
Plate Iron,
Shingle Strips,
Anvils,
Crow Bars,
Harrow Teeth,
Horse Shoes,
Tool Steel,
Machinery Steel,
Spring Steel,
Plow Steel,
Frog Steel,
Lay Steel,
Sheet Steel,
Saw Steel,
Safe Steel,
Hammer Steel,
Tire Steel,
Blister Steel.

FORT PITT IRON and STEEL WORKS,

PITTSBURG, PA.

Tensil Strength of Bar Iron, 60,000 lbs. to the Square inch.
' Machinery Steel, 100,000 " " "
" Tool " 140,000 " " "

Clamps.

Champlin & Rogers........Chicago, Ill........See Page 192

Claw and Tamping Bars.

Skinner & Gifford Mfg. Co........Dunkirk, N. Y........See Page 140

Clocks.

Ansonia Brass & Copper Co........New York........See Page 122
Vose, Dinsmore & Co........" "........" 60

Coach Colors.

Edward Smith & Co........New York........See Page 218

Coach Oil.

Eclipse Lubric Oil Co........New York........See Page Facing Title
F. S. Pease........Buffalo, N. Y........See Page 100

Coach Screws.

G. B. Walbridge........New York........See Page 130
Lewis Oliver & Phillips; H. B. Newhall, Agt.. " "........" 166
Providence Tool Co.; " " ".. " "........" 166

Coach Varnish.

Edward Smith & Co........New York........See Page 218
Valentine & Co........" "........" 198

Coal.

E. L. Hedstrom........Buffalo, N. Y........See Page 202
Rhodes & Bradley........Chicago, Ill........" 192
Rogers & Co........" "........" 144

WEIGHT OF FLAT IRON.—Continued.

Weight of Running Foot in Pounds.

Thickness in Inches.

Width in Ins.	7/16	1/2	5/8	3/4	7/8	1
1	1.46	1.67	2.08	2.5	2.92	3.33
1/8	1.64	1.87	2.34	2.81	3.28	3.75
1/4	1.82	2.08	2.6	3.12	3.65	4.17
3/8	2.01	2.29	2.86	3.44	4.01	4.58
1/2	2.19	2.5	3.12	3.75	4.37	5.
5/8	2.37	2.71	3.38	4.06	4.74	5.42
3/4	2.55	2.92	3 64	4.37	5.1	5.83
7/8	2.73	3.12	3.9	4.69	5.47	6.25
2	2.92	3.33	4.16	5.	5.83	6.67
1/8	3.1	3.54	4.43	5.31	6.2	7.08
1/4	3.28	3.75	4.69	5.62	6.56	7.5
3/8	3.46	3.96	4.95	5.94	6.93	7.92
1/2	3.65	4.17	5.21	6.25	7.29	8.33
5/8	3.83	4.37	5.47	6.56	7.66	8.75
3/4	4.01	4.58	5.73	6.88	8.02	9.17
7/8	4.19	4.79	5.99	7.19	8.39	9.58
3	4.37	5.	6.25	7.5	8.75	10.
1/8	4.56	5.21	6.51	7.82	9.12	10.42
1/4	4.74	5.42	6.77	8.12	9.48	10.83
3/8	4.92	5.62	7.03	8.44	9.84	11.25
1/2	5.1	5.83	7.29	8.75	10.21	11.67
5/8	5.29	6.04	7.55	9.07	10.59	12.08
3/4	5.47	6.25	7.81	9.37	10 93	12.5
7/8	5.65	6.46	8.07	9.68	11.3	12.92
4	5.83	6.67	8.33	10.	11.65	13.33
1/8	6.02	6.87	8.59	10.3	12.04	13.75
1/4	6.2	7.08	8.85	10.62	12.4	14.15
3/8	6.38	7.29	9.11	10.93	12.75	14.59
1/2	6.56	7.5	9.37	11.25	13.12	15.00
5/8	6.74	7.71	9.64	11.55	13.5	15.42
3/4	6.93	7.92	9.89	11.87	13.85	15.83
7/8	7.11	8.12	10.15	12.2	14.22	16.25
5	7.29	8.33	10.42	12.5	14.59	16.65
1/8	7.48	8.54	10.69	12.8	14.95	17.09
1/4	7.66	8.75	10.93	13.13	15.3	17.5

Coal Facing.

Rogers & Co....Chicago, Ill....See Page 144

Coal Hods and Vases.

Sidney Shepard & Co....Buffalo, N. Y....See Page 196

Coal Miners' Supplies.

Hart, Ball & Hart....Buffalo, N. Y....See Page 136

Coal Washing Machinery.

The Atlas Works....Pittsburgh, Pa....See Page 172

Coiled Chain.

De Grauw, Aymar & Co....New York....See Page 112
Hermann Boker & Co.... " " ....See Pages 152, 154

Coiled Springs.

C. Van Ness....New York....See Page 70

Coking Machinery.

The Atlas Works....Pittsburgh, Pa....See Page 172

Cold Rolled Piston Rods.

Jones & Laughlins....Pittsburgh, Pa....See Page 32

Colors—Dry.

C. E. Hecht....Easton, Pa....See Page 158
Edward Smith & Co....New York.... " 218
Jno. W. Masury & Son.... " " .... " 214

WEIGHT OF FLAT IRON.—CONTINUED.

Weight of Running Foot in Pounds.

Width in Ins.	Thickness in Inches.					
	7/16	1/2	5/8	3/4	7/8	1
5 3/8	7.84	8.96	11.2	13.43	15.68	17.92
1/2	8.02	9.17	11.45	13.75	16.03	18.33
5/8	8.2	9.37	11.72	14.07	16.4	18.75
3/4	8.39	9.58	11.99	14.37	16.77	19.15
7/8	8.57	9.79	12.25	14.7	17.13	19.58
6	8.75	10.	12.5	15,	17.5	20.
1/8	8.93	10.2	12.77	15.3	17.85	20.42
1/4	9.11	10.42	13.02	15.62	18.23	20.83
3/8	9.3	10.63	13.29	15.93	18.6	21.25
1/2	9.48	10.83	13.53	16.25	18.97	21.65
5/8	9.67	11.03	13.81	16.57	19.33	22.08
3/4	9.84	11.25	14.05	16.87	19.7	22.5
7/8	10.02	11.45	14.32	17.19	20.03	22.92
7	10.2	11.65	14.59	17.5	20.42	23.33
1/4	10.59	12.09	15.1	18.13	21.15	24.15
1/2	10.93	12.5	15.62	18.73	21.85	25.
3/4	11.31	12.92	16.15	19.39	22.62	25.83
8	11.66	13.33	16.65	20.	23.33	26.65
1/4	12.03	13.75	17.18	20.6	24.05	27.5
1/2	12.4	14.17	17.7	21.25	24.8	28.33
3/4	12.76	14.58	18.23	21.89	25.52	29.17
9	13.12	15.	18.75	22.5	26.23	30.
1/4	13.5	15.42	19.27	23.12	26.98	30.83
1/2	13.85	15.83	19.78	23.73	27.7	31.67
3/4	14.2	16.25	20.32	24.35	28.42	32.5
10	14.59	16.65	20.82	25.	29.15	33.33
1/4	14.93	17.08	21.33	25.62	29.88	34.17
1/2	15.3	17.5	21.89	26.25	30.62	35.
3/4	15.67	17.92	22.4	26.85	31.33	35.83
11	16.03	18.33	22.9	27.5	32.08	36.65
1/4	16.4	18.75	23.43	28.12	32.8	37.5
1/2	16.75	19.15	23.93	28.73	33.52	38.33
3/4	17.13	19.59	24.49	29.35	34.25	39.15
12	17.5	20.	25.	30.	35.	40.

Colors, for Freight and Coal Cars.

C E. Hecht.......... Easton, Pa.......... See Page 158
Jno. W. Masury & Son.......... New York.......... " 214

Colors—In Oil.

C. E. Hecht.......... Easton, Pa.......... See Page 158
Jno. W. Masury & Son.......... New York.......... " 214

Commission Merchants.

C. W. Matthews.......... Philadelphia, Pa.......... See Page 38
Jno. W. Quincy.......... New York.......... " 178
Winans & Co.......... " " .... " 106

Compasses.

F. Eckel.......... New York.......... See Page 76
Kuebler & Seelhorst.......... Philadelphia, Pa.......... " 216
Wm. J. Young & Sons.......... " " .... " 138

Composition Castings.

Philadelphia Smelting Co.......... Philadelphia, Pa.......... See Page 134

Condensers.

A. Carr.......... New York.......... See Page 244

Conductors' Punches.

Allen, Lane & Scott.......... Philadelphia, Pa.......... See Page 90
Geo. H. Crain & Co.......... Chicago, Ill.......... " 124
L. W. Pond.......... Worcester, Mass.......... " 158
Vose, Dinsmore & Co.......... New York.......... " 60

Contractors.

Geo. H. Crain & Co.......... Chicago, Ill.......... See Page 124
Joseph Churchyard.......... Buffalo, N. Y.......... " 160
Phillipsburg Manufacturing Co.......... New York.......... " 22

Contractors' Machinery.

Speedwell Iron Works.......... New York.......... See Page 122

LAP WELDED AMERICAN CHARCOAL IRON BOILER TUBES.

Table of Standard Sizes.—Morris, Tasker & Co.

External Diameter.	External Circumference.	Internal Diameter.	Internal Circumference.	Thickness.	Length of Pipe per square foot of inside surface.	Length of Pipe per square foot of outside surface.	Internal Area.	External Area.	Weight, per foot.
Ins.	Ins.	Ins.	Ins.	Ins.	Feet.	Feet.	Ins.	Ins.	lbs.
1	3.142	0.856	2.689	0.072	4.460	3.819	0.575	0.785	0.708
1¼	3.927	1.106	3.474	0.072	3.455	3.056	0.960	1.227	0.9
1½	4.712	1.334	4.191	0.083	2.863	2.547	1.396	1.767	1.250
1¾	5.598	1.560	4.901	0.095	2.448	2.183	1 911	2.405	1.665
2	6.283	1.804	5.667	0.098	2.118	1.909	2.556	3.142	1.981
2¼	7.069	2.054	6.484	0.098	1.850	1.693	3.314	3.976	2.238
2½	7.854	2.283	7.172	0.109	1.673	1.528	4.094	4.909	2.755
2¾	8.639	2.533	7.957	0.109	1.508	1.390	5.039	5.940	3.045
3	9.425	2.783	8.743	0.109	1.373	1.273	6.083	7.069	3.333
3¼	10.210	3.012	9.462	0.119	1.268	1.175	7.125	8.296	3.958
3½	10.995	3.262	10.248	0.119	1.171	1.091	8.357	9.621	4.272
3¾	11.781	3.512	11.033	0.119	1.088	1.018	9.687	11.045	4.590
4	12.566	3.741	11.753	0.130	1.023	0.955	10.992	12.566	5.320
4½	14.137	4.241	13.323	0.130	0.901	0.849	14.126	15.904	6.010
5	15.708	4.72	14.818	0.140	0.809	0.764	17.497	19.635	7.226
6	18.849	5.699	17.904	0.151	0.670	0.637	25.509	28.274	9.346
7	21.991	6.657	2 .914	0.172	0.574	0.545	34.805	38.484	12.435
8	25.132	7.636	23.989	0.182	0.500	0.478	45.795	50.265	15.109
9	28.274	8.615	27.055	0.193	0.444	0.424	58.291	63.617	18.002
10	31.416	9.573	30.074	0.214	0.399	0.382	71.975	78.540	22.19

Contractors' Supplies.

Greene & Randolph.....New York.....See Page 54
Pratt & Co.....Buffalo, N. Y..... " 132

Contractors' Tools.

Jno. Bayliss.....New York.....See Page 76

Copper.

Ansonia Brass & Copper Co.....New York.....See Page 122
Bruce & Cook..... " " ..... " 112
Chas. W. Matthews.....Philadelphia, Pa..... " 38
Geo. Worthington & Co.....Cleveland, Ohio..... " 100
Holmes & Lissberger.....New York..... " 84
Philadelphia Smelting Co.....Philadelphia, Pa..... " 134
Sidney Shepard & Co.....Buffalo, N. Y..... " 196

Copper—Bottoms.

Holmes & Lissberger.....New York.....See Page 84

Copper—Ingot.

Lucius Hart & Co.....New York.....See Page 56
Jno. W. Quincy..... " " ..... " 178

Copper—Locomotive Sheets.

Ansonia Brass & Copper Co.....New York.....See Page 122
Holmes & Lissberger..... " " ..... " 84

Copper Nails.

Ansonia Brass & Copper Co.....New York.....See Page 122

Copper Rivets and Burs.

Ansonia Brass & Copper Co.....New York.....See Page 122
Holmes & Lissberger..... " " ..... " 84

Copper Tubing.

Ansonia Brass & Copper Co.....New York.....See Page 122
Holmes & Lissberger..... " " ..... " 84

PAINTER'S PUTTY.

Spanish whiting, pulverized......	81.6	Made into a stiff paste. If not intended for immediate use, raw oil should be used.
Boiled Oil......	20.4	

One pound of putty for stopping every 20 yards.

GLAZIER'S PUTTY.

Whiting, 70 pounds; boiled oil, 30 pounds; water, 2 gallons. Mix. If too thin, add more whiting; if too thick, add more oil.

TO SOFTEN PUTTY.

To remove old putty from broken windows, dip a small brush in nitro-muriatic acid or caustic soda (concentrated lye), and with it annoint or paint over the dry putty that adheres to the broken glass and frames of your windows; after an hours interval, the putty will have become so soft as to be easily removable.

WEIGHT OF ONE FOOT OF COPPER RODS.

Diameter in inches.	Pounds.	Diameter in inches.	Pounds.	Diameter in inches.	Pounds.
¼	.189	1⅛	3.831	2⅜	17.075
5-16	.295	3-16	4.269	½	18.916
⅜	.425	¼	4.730	⅝	20.856
7-16	.579	5-16	5.214	¾	22.891
½	.756	⅜	5.730	⅞	25.019
9-16	.958	7-16	6.254	3.	27.243
⅝	1.182	½	6.811	⅛	29.560
11-16	1.430	9-16	7.389	¼	31.972
¾	1.702	⅝	7.993	⅜	34.481
13-16	1.998	¾	9.270	½	37.080
⅞	2.317	⅞	10.642	⅝	39.777
15-16	2.660	2.	12.108	¾	42.568
1.	3.027	⅛	13.667	⅞	45.455
1-16	3.417	¼	15.325	4.	48.433

Copper Wire.

Ansonia Brass & Copper Co	New York	See Page 122

Copper Wire—Insulated.

Charles T. Chester	New York	See Page 212
Geo. W. Mowbray	North Adams, Mass	" 182

Copying Presses.

T. Shriver & Co	New York	See Page 194

Cordage.

De Grauw, Aymar & Co	New York	See Page 112
Sidney Shepard & Co	Buffalo, N. Y	" 196

Corrugated Iron.

John Merry & Co	New York	See Page 48
Marshall Lefferts, Jr	" "	" 44
Noyes & Wines	" "	" 264
Tyng & Co	New York	" 260

Cotter Drills.

Wm. Sellers & Co	Philadelphia, Pa	See Page 64

Cotton Waste.

J. J. Walworth	Chicago, Ill	See Page 124
L. G. Tillotson & Co	New York	" 94
Radley & McAlister Manufacturing Co	" "	" 250
T. B. Bickerton & Co	Philadelphia, Pa	" 138
Vose, Dinsmore & Co	New York	" 60

Couplings.

A. Fulton's Son & Co	Pittsburgh, Pa	See Page 116
Darrow, Turner & Co	New York	" 62
Radley & McAlister Manufacturing Co	" "	" 250
Vose, Dinsmore & Co	" "	" 60

Covered Wires—Silk and Cotton.

Bishop Gutta Percha Works	New York	See Page 34
C. Thompson	Philadelphia, Pa	" 222
Charles T. Chester	New York	" 212
Geo. H. Bliss & Co	Chicago, Ill	" 164
Leclanche Battery Co	New York	" 96

SURFACE HARDENING OF CAST-IRON.

The wearing of cast-iron surfaces exposed to sliding friction can be almost wholly prevented by tempering the surface with a mixture of 21 1-10 pints of water, 30¾ pounds of sulphuric acid, and 1,003 grains of nitric acid. The article should be heated to a cherry-red, and protected from the oxidizing effect of currents of air by a sheet-iron box. The process is especially adapted to the hardening of bearings of axles, which, while much cheaper than those of the usual alloy, will, when regularly ubricated, last as long, even when there is great rapidity of motion.

RIVETED IRON PIPES.

Weight per Running Foot.

These weights include laps for riveting and calking, but not the rivets, the weight of which depends upon the sizes and number used.

Bore. Inch.	Thickness of Metal, Inch. 1/8	3/16	1/4
5	7.15	10.7	14.23
½	7.8	11.71	15.61
6	8.45	12.6	16.85
½	9.05	13.61	18.15
7	9.75	14.65	19.55
½	10.45	15.7	20.85
8	11.15	16.65	22.25
½	11.85	17.6	23.55
9	12.5	18.75	25.
½	13.15	19.75	26.25
10	13.96	20.8	27.7
½	14.45	21.75	29.1

Bore. Inch.	Thickness of Metal, Inch. 3/16	1/4	5/16
11	22.75	30.45	38.15
12	24.8	33.	41.25
13	26.75	35.75	44.55
14	28.65	38.45	47.1
15	30.83	41.	51.45
16	32.94	43.75	54.75
17	34.85	46.45	58.1
18	36.3	49.1	61.4
19	38.	51.75	64.7
20	41.2	55.6	68.

Baldwin Locomotive Works

ALL WORK ACCURATELY FITTED TO GAUGES,

AND THOROUGHLY INTERCHANGEABLE,

Plan, Materials, Workmanship, Finish and Efficiency

FULLY GUARANTEED

M. BAIRD.	CHAS. T. PARRY.	WM. P. HENSZEY.
GEO. BURNHAM.	EDW'D H. WILLIAMS.	EDW'D LONGSTRETH.

Cranes.

Wm. Sellers & Co....Philadelphia, Pa....See Page 64
Wm. B. Bement & S n.... " " .... " 168

Crestings.

J. D. West & Co....New York....See Page 56

Crossing and Frogs.

Anderson & Woods....Pittsburgh, Pa....See Page 150
The Atlas Works.... " " .... " 172

Crow-Bars—Iron and Steel.

Anderson & Woods....Pittsburgh, Pa....See Page 150
G. B. Walbridge....New York.... " 130
Hermann Boker & Co.... " " ....See Pages 152, 154
Providence Tool Co; H. B. Newhall, Agent.... " " ....See Page 166
Reese, Graff & Woods....Pittsburgh, Pa.... " 68

Crucibles.

Du Plaine & Reeves....Philadelphia, Pa....See Page 102
Van Tuyl Manufacturing Co....New York.... " 242

Crude Gutta Percha for Cement.

Bishop Gutta Percha Works....New York....See Page 34

Cupola Linings.

Hall & Sons....Buffalo, N. Y....See Page 254
Maurer & Weber....New York.... " 228
Philip Neukumet....Philadelphia, Pa.... " 104
Star Fire Brick Co....Pittsburgh, Pa.... " 232

DRIVING WHEELS.

Number of Revolutions per Mile.

Diameter of Wheel	2 ft.	2½ ft.	3 ft.	3½ ft.	4 ft.	4½ ft.	5 ft.
Revolutions per Mile	840	672	560	480	420	373	336

Diameter of Wheel	5½ ft.	6 ft.	6½ ft.	7 ft.	8 ft.	9 ft.	10 ft.
Revolutions per Mile	305½	280	258½	240	210	187	168

VELOCITY AND FORCE OF THE WIND.

Description.	Miles per hour.	Feet per minute.	Feet per second.	Force in lbs. per sq. foot.
Hardly perceptible	1	88	1.47	.005
Just perceptible	2	176	2.93	.020
	3	264	4.4	.044
Gentle breeze	4	352	5.87	.079
	5	440	7.33	.123
Pleasant breeze	10	880	14.67	.492
	15	1320	22	1.107
Brisk gale	20	1760	29.3	1.968
	25	2200	36.6	3.075
High wind	30	2640	44	4.428
	35	3080	51.3	6.027
Very high wind	40	3520	58.6	7.872
	45	3960	66.0	9.963
Storm	50	4400	73.3	12.300
Great Storm	60	5280	88.0	17.712
	70	6160	102.7	24.108
Hurricane	80	7040	117.3	31.488
	100	8800	146.6	49.200

PAINT FOR TARPAULINS.

First. Olive—Liquid, Olive color.... 100.
Beeswax.... 6.
Spirits turpentine.... 6.
Dissolve the beeswax in spirits of turpentine, with a gentle heat, and mix the paint warm.

Second. Add 12 ounces of beeswax to one gallon of linseed oil, boil it two hours; prime the cloth with this mixture, and use it in place of boiled oil, for mixing the paint.

Lathes,
Planers,
Slotters,
Shapers,
Boring Mills,
Bolt Cutters,
Bolt Headers,
Rivet Machines,
Milling Machines,
Screw Machines,
Nut Tappers,
Gear Cutters,
Gang Drills,
Axle Lathes,

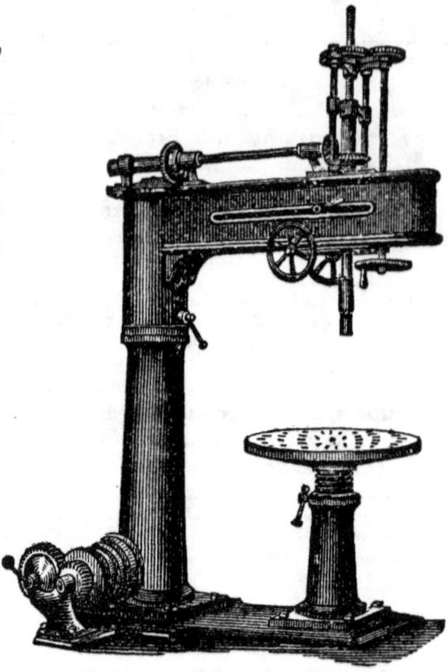

Car Wheel Boring Mills,
Wheel Presses,
Scrap Shears,
Punch and Shears,
Vertical Drills,
Cutting Off Machines,
Radial Drills,
Steam Pumps,
Portable Hoisters
Portable Drills,
Agricultural Engines,
Shafting, Hangers,
Pulleys, Belting,
&c., &c.

ISAAC H. SHEARMAN,

132 North Third St., and 301 Cherry St.,

PHILADELPHIA,

MACHINERY COMMISSION HOUSE,

Steam Engines, Boilers,

AND

MACHINISTS' TOOLS,

SPECIALTIES.

Warren's Shaper. Schlenker's Bolt Cutter.

Durrell's Nut Tapper *(Taps 7 Nuts at Once.)*

Barr's Radial Drill.

Heavy Lathes, Adapted to Railroad Work.

Cupolas.

Wm. B. Bement & Son.............................Philadelphia, Pa.............See Page 168
Wm. Sellers & Co.. " " .... " 64

Cutlery.

England & Bindley......................................Pittsburgh, Pa ....See Page 162
G. B. Walbridge...New York .... " 130
Hermann Boker & Co................................. " " ....See Pages 152, 154
Pratt & Co...Buffalo, N. Y..................See Page 132

Cutting Nippers.

Biddle Manufacturing Co..............................New York......................See Page 12

Cylinders.

The Atlas Works....... ....Pittsburgh, Pa.................See Page 172

Cylinder Boring and Facing Machines.

Wm. Sellers & Co............. ....Philadelphia, Pa............See Page 64

Dating Machines.

Allen, Lane & Scott.....................................Philadelphia, Pa. ....See Page 90
Mathiesen & Schrader..................................New York..................... " 82

Depot Scales.

Buffalo Scale Company................................Buffalo, N. Y..................See Page 40
Jones Scale Works.......................................Binghamton, N. Y......... " 92

Derricks.

Skinner & Gifford Manufacturing Co............Dunkirk, N. Y...............See Page 140

STEAM.

THE ELASTIC FORCE OF STEAM AND CORRESPONDING TEMPERATURE OF THE WATER WITH WHICH IT IS IN CONTACT.

(*From Haslett.*)

Pressure on a Square Inch. lbs.	Elastic Force in Inches of Mercury.	Temperature of Water in Degrees of Fahrenheit.	Volume of Steam compared with the Volume of Water.	Pressure on a Square Inch. lbs.	Elastic Force in Inches of Mercury.	Temperature of Water in Degrees of Fahrenheit.	Volume of Steam compared with the Volume of Water.
14.7	30.00	212.0	1700	68	138.72	304.4	419
15	30.60	212.8	1669	70	142.80	306.4	408
16	32.64	216.3	1573	72	146.88	308.4	398
18	36.72	222.7	1411	74	150 96	310.3	388
20	40.80	228.5	1281	76	155.06	312.2	379
22	44.88	233.8	1174	78	159.14	314.0	370
24	48.96	23 .7	1084	80	163.22	315.8	362
26	53.04	243.3	1007	82	167.30	317.6	354
28	57.12	247.6	941	84	171.38	319.3	346
30	61.21	251.6	883	86	175.46	321.0	339
32	65.28	255.5	833	88	179.54	322.6	332
34	69.36	259.1	788	90	183.62	324.3	325
36	73.44	262.6	748	92	187.70	325.9	319
38	77.52	265.9	712	94	191.78	327.5	313
40	81.60	269.1	679	96	195.86	329.0	307
42	85.68	272.1	649	98	199.92	330.5	301
44	89.76	275.0	622	100	204.01	332.0	295
46	93.84	277.8	598	110	224.40	339.2	271
48	97.92	280.5	575	120	244.82	345.8	251
50	102.00	283.2	554	130	265.23	352.1	233
52	106.08	285.7	534	140	285.61	357.9	218
54	110.16	288.1	516	150	306.03	363.4	205
56	114 24	290.5	500	160	326.42	368.7	193
58	118.32	292.9	484	170	346.80	373.6	183
60	122.40	295.6	470	180	367.25	378.4	174
62	126.48	298.1	456	190	387.61	382.9	166
64	130.56	300.3	443	200	408.04	387.3	158
66	134.64	302.4	431				

Desks and Office Furniture.

A. L. Hale & Bro .... Chicago, Ill .... See Page 206
Joseph Churchyard .... Buffalo, N. Y .... " 160

Dies.

Champlin & Rogers .... Chicago, Ill .... See Page 192
J. J. Walworth .... " " .... " 124
Mathiesen & Schrader .... New York .... " 82

Door Knobs.

Romer & Co .... Newark, N. J .... See Page 201

Door Machinery.

C. B. Rogers & Co .... New York .... See Page 210
J. A. Fay & Co .... Cincinnati, Ohio .... " 206
Richards, London & Kelley .... Philadelphia, Pa .... " 66

Drawing Room Car Hassocks, &c.

George Drake Smith .... Philadelphia, Pa .... See age 246
New York Ottoman & Hassock Co .... New York .... " 236

Drill Grinding Machines.

Wm. Sellers & Co .... Philadelphia, Pa .... See Page 64

STRENGTH AND EQUIVALENTS OF FLAT ROPES.

Manchester (England) List.

Bre king Strain. Tons.	Working Load. Cwt.	HEMP. Circumference. Inches.	IRON WIRE. Circumfereuce. Inches.	STEEL WIRE. Circumference. Inches.
16	32	3 by 1	1⅞ by ½	
20	40	4 " 1⅛	2 " ⅝	
22	48	5 " 1¼	2¾ " ½	
24	56	5¼ " 1½	2⅞ " ⅝	1⅞ by ⅝
26	64	5½ " 1⅜	3 " ⅝	2 " ⅝
27	72	5¾ " 1½	3¼ " ⅝	
28	80	6 " 1½	3½ " ¾	2¾ " ½
32	88	6½ " 1⅝	3¾ " ¾	
36	96	7 " 1⅞	4 " ¾	2⅞ " ⅝
40	108	7½ " 2[illegible]	4¼ " ⅞	3 " ⅝
46	120	8½ " 2¼	4½ " ⅞	3¼ " ⅝
58	145	10 " 2½	5 " ⅞	3¼ " ¾
70	175	11½ " 2¾	5½ " 1	4¼ " ⅞
85	210	13 " 3	6 " 1	4⅝ " ⅞

WEIGHT OF FLAT ROPES.

Manchester (England) List.

HEMP. Size in Inches.	HEMP. Weight per Fathom. Lbs.	IRON WIRE. Size in Inches.	IRON WIRE. Weight per Fathom. Lbs.	STEEL WIRE. Size in Inches.	STEEL WIRE. Weight per Fathom. Lbs.
3 by 1	16	1⅞ by ½	8	1⅞ by ½	8
4 " 1⅛	20	2 " ⅝	10	2 " ⅝	10
5 " 1¼	22	2¾ " ½	12	2¾ " ½	12
5¼ " 1¼	24	2⅞ " ½	14	2⅞ " ½	12
5½ " 1⅜	26	3 " ⅝	16	3 " ⅝	16
5¾ " 1½	27	3¼ " ⅝	18	3¼ " ⅝	18
6 " 1½	28	3½ " ¾	20	3¾ " ¾	24
6½ " 1⅝	32	3¾ " ¾	22	4¼ " ¾	28
7 " 1⅞	36	4 " ¾	24	4⅝ " ⅞	34
7½ " 2⅛	40	4¼ " ⅞	28		
8½ " 2¼	45	4½ " ⅞	30		
10 " 2½	63	5 " ⅞	39		
11½ " 2¾	78	5½ " 1	48		
13 " 3	98	6 " 1	59		

Drills.

Barwick Wrench Co....................Boston, Mass....................See Page 60
L. W. Pond....................Worcester, Mass....................“ 158
Rand & Waring Drill & Compressor Co....................New York....................See Front Page.
See also Drilling Machines, Tools, &c.

WIND MILLS.

Deduction from Velocities varying from 4 to 9 Feet per Second. (Haswell.)

1. The velocity of wind-mill sails, so as to produce a maximum effect, is nearly as the velocity of the wind, their shape and position being the same.

2. The load at the maximum is nearly, but somewhat less than, as the square of the velocity of the wind, the shape and position of the sails being the same.

3. The effects of the same sails, at a maximum, are nearly, but somewhat less than, as the cubes of the velocity of the wind.

4. The load of the same sails, at the maximum is nearly as the squares, and their effect as the cubes of their number of turns in a given time.

5. When sails are loaded so as to produce a maximum effect at a given velocity, and the velocity of the wind increases, the load continuing the same—*First*, the increase of effect, when the increase of the velocity of the wind is small, will be nearly as the squares of those velocities; *Secondly*, when the velocity of the wind is double, the effects will be nearly as 10 to 27½; *Thirdly*, when the velocities compared are more than double of that when the given load produces a maximum, the effects increase in nearly the same ratio of the velocity of the wind.

6. In sails where the figure and position are similar, and the velocity of the wind the same, the number of revolutions in a given time will be reciprocally as the radius or length of the sail.

7. The load, at a maximum, which sails of a similar figure and position will overcome at a given distance from the centre of motion is as the cube of the radius.

8. The effects of sails of similar figure and position are as the square of the radius.

9. The velocity of the extremities of Dutch sails, as well as of the enlarged sails, in all their usual positions when unloaded, or even loaded to a maximum, is considerably greater than that of the wind.

Rule for the Angles of the Sails. (Molesworth.)

A = Angle of the sail with the plane of motion at any part of the sail.
R = Total radius of sail in feet.
D = Distance of any part of the sail from the axis.

$$A = 23° - \frac{18 D^2}{R^2}$$

If the radius of the wind mill sails be divided into six equal parts, the angles at each of those parts will be as follows, reckoning from the axis:

Distance from axis	1-6	2-6	3-6	4-6	5-6	tip
Angle of sail with axis	67½°	69°	71½°	75°	79½°	85°
Angle of sail with plane of motion	22½	21	18½	15	10½	5

Axis of shaft of wind mill with horizon.................... = 8° on level ground.
“ “ “ “ .... = 15° on high exposed positions.
Breadth of whip at axis.................... = 1-30th length of whip.
Depth “ “ .... = 1-40th “
Breadth of whip at tip.................... = 1-60th “
Depth “ “ .... = 1-80th “
Width of sail “ .... = 1-3d “
Divided by the whip in the proportion of 5 to 3, the narrow portion being nearest to the wind.
Width of sail at axis.................... = 1-5th length of whip.
Distance of sail from axis.................... = 1-7th “
Cross-bars from 16 to 18 inches apart.

Horse-Power and Sail-Area of Wind Mills.

HP = Horse-power.
V = Velocity of wind in feet per second.
A = Total area of sails in square feet.
N = Number of sails.

$$A = \frac{HP\ 1080000}{V^3}.$$

$$HP = \frac{A\ V^3}{1080000}.$$

Area of each sail = A ÷ N. Velocity of tips of sails = 2.6 V, nearly.

Drilling Machines.

Buffalo Machinery Agency..........Buffalo, N. Y..........See Page 160
England & Bindley..........Pittsburgh, Pa..........“ 162
Isaac H. Shearman..........Philadelphia, Pa..........“ 80
Rand & Waring Drill & Compressor Co..........New York..........See Front Page
The Pratt & Whitney Co..........Hartford, Conn..........See Page 142
Thorne, De Haven & Co..........Philadelphia, Pa..........“ 38
Wm. B. Bement & Son..........“ “..........“ 168

Drill Presses.

Phelps & Sanger..........Cleveland, Ohio..........See Page 120
Wm. Sellers & Co..........Philadelphia, Pa..........“ 64

Driving Wheel Lathes.

Wm. Sellers & Co..........Philadelphia, Pa..........See Page 64

Drop Presses.

Winans & Co..........New York..........See Page 106

Dryers.

Brooklyn White Lead Co..........New York..........See Page 268
Jno. W. Masury & Son..........“ “..........“ 214
Wetherill & Bro..........Philadelphia, Pa..........“ 52

Edge Tools.

G. B. Walbridge..........New York..........See Page 130
Hermann Boker & Co..........“ “..........See Pages 152, 154

Electrical Engineers.

Charles T. Chester..........New York..........See Page 212
M. A. Buell..........Cleveland, Ohio..........“ 46
Towle & Unger Manufacturing Co..........New York..........“ 186

Electric Alarms.

Charles T. Chester..........New York..........See Page 212
M. A. Buell..........Cleveland, Ohio..........“ 46
Towle & Unger Mfg. Co..........New York..........“ 186

STRENGTH AND EQUIVALENTS OF ROUND ROPES.

Manchester (England) List.

Breaking Strain. Tons.	Working Load. Cwt.	Hemp. Circumference. Inches.	Iron Wire. Circumference. Inches.	Steel Wire. Circumference Inches.
¾	1½	1½	¾	
1	3	1¾	⅞	
1½	4	2	1⅛	¾
2	6	2¼	1¼	⅞
2½	8	2½	1⅜	1⅛
3	10	2¾	1½	1¼
4	15	3	1¾	1⅜
5	18	3½	1⅞	1½
6	20	3¾	2	
6½	22	4	2⅛	
7	24	4½	2¼	1¾
8	27	4¾	2⅜	
9	28	5	2½	1⅞
10	34	5½	2⅝	2
11	36	6	2¾	2⅛
13	42	6½	2⅞	2¼
14	45	7	3	2⅜
15	50	7½	3⅛	2½
17	57	8	3⅜	2⅝
19	66	8½	3½	2¾
21	70	9	3⅝	2⅞
23	78	9½	3¾	3
25	84	10	4	3⅛
29	96	11	4¼	3⅜
34	108	11½	4⅝	3½
36	120	12	5	3⅝
41	135	12½	5½	4
48	160	13	6	4¼

GEO. E. NEWCOMB. *JAS. W. NEWCOMB.* *JOHN H. NEWCOMB*

NEWCOMB BRO'S. SONS,

MANUFACTURERS OF

SMITHS', MOULDERS' and HAND

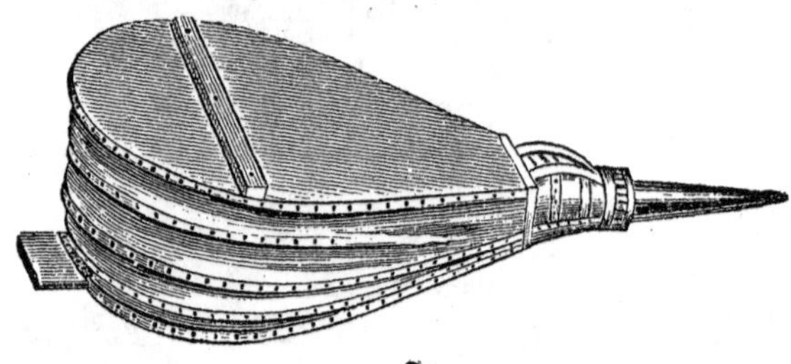

ALSO THE SOLE MANUFACTURERS

OF

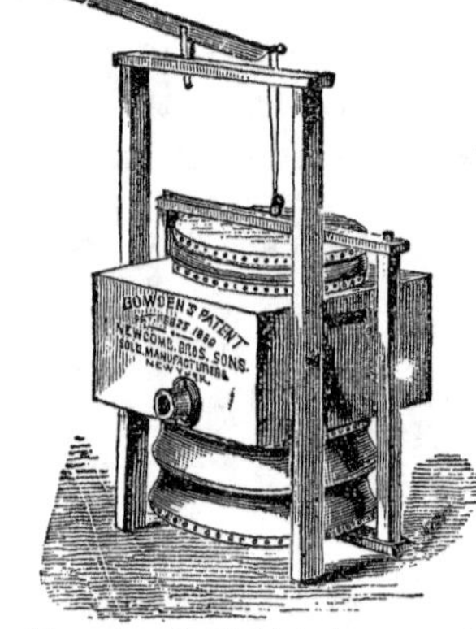

BOWDEN'S PATENT

TRIPLE-ACTION

FORGE BELLOWS.

These Bellows are superior to all others now in use.

They can be used either in single or double action, for light or heavy work as required. They maintain a steady and continuous blast, the air being forced into the chest both at the up and down stroke. The strength of blast obtained is double that of the ordinary bellows of the same dimensions. They work easily. While the blast is double, the power required to work them is no more than that of a common bellows. Being fully rigged and geared, ready for work, and easily moved from place to place, they are very convenient for railroad, mining, and other purposes where portable bellows are required. They economize space, occupying only half the room of an ordinary bellows. These bellows come to hand at once. A good blast is obtained at the first stroke, and no time or labor is lost, as in the old single-action bellows, in getting up a good blast. They can be made of any size and shape, to suit all requirements. They are the cheapest bellows in the world. With all the advantages of double-action, they cost no more, power for power, than the common single action of same quality.

☞ *For further particulars send for descriptive circular and price list.*

586 WATER ST., near Montgomery, NEW YORK.

STATE RIGHTS FOR SALE.

Canal Street, East Broadway and South Street Cars cross Montgomery Street.

Electric Apparatus for Mining.

Geo. W. Mowbray	North Adams, Mass	See Page 182
Laflin & Rand Powder Co	New York	See Last Page

Electric Bells.

Charles T. Chester	NewYork	See Page 212
Geo. H. Bliss & Co	Chicago, Ill	" 164
Leclanche Battery Co	New York	" 96

Electric Instruments.

Charles T. Chester	New York	See Page 212
F. L. Pope & Co	" "	" 184
Geo. H. Bliss & Co	Chicago, Ill	" 164
Leclanche Battery Co	New York	" 96
L. G. Tillotson & Co	" "	" 94
M. A. Buell	Cleveland, Ohio	" 46
Merchants' Manufacturing & Construction Co.	New York	" 18
Towle & Unger Manufacturing Co	" "	" 186

Electric Machinery.

Charles T. Chester	New York	See Page 212
F. L. Pope & Co	" "	" 184
Geo. H. Bliss & Co	Chicago, Ill	" 164
Leclanche Battery Co	New York	" 96
L. G. Tillotson & Co	" "	" 94
M. A. Buell	Cleveland, Ohio	" 46
Merchants' Manufacturing & Construction Co.	New York	" 18
Towle & Unger Manufacturing Co	" "	" 186

WEIGHT OF ROUND ROPES.

Manchester (England) List.

Hemp.		Iron Wire.		Steel Wire,	
Circumference. Inches.	Weight per fathom. Lbs.	Circumference. Inches.	Weight per fathom. Lbs.	Circumference. Inches.	Weight per fathom. Lbs.
1½	¾	¾	½	¾	½
1¾	1	⅞	¾	⅞	¾
2	1½	1	⅞	1⅛	1
2¼	2	1⅛	1	1¼	1¼
2½	2½	1¼	1¼	1⅜	1½
2¾	3	1⅜	1½	1½	1¾
3	3½	1½	1¾	1¾	2½
3½	4	1¾	2½	1⅞	3
3¾	4½	1⅞	3	2	3¼
4	5	2	3¼	2⅛	3¾
4½	6	2⅛	3¾	2¼	4
4¾	6½	2¼	4	2⅜	4¼
5	7	2⅜	4½	2½	4¾
5½	8	2½	4¾	2⅝	5½
6	9	2⅝	5½	2¾	6
6½	10	2¾	6	2⅞	7
7	12	2⅞	7	3	7½
7½	14	3	7½	3⅛	8¼
8	16	3⅛	8¼	3⅜	9¼
8½	20	3⅜	9½	3½	11
9	22	3½	11	3⅝	11¾
9½	25	3¾	13	4	14
10	28	4	14	4¼	16
11	29	4¼	16		
11½	30	4⅝	18		
12	32	5	20		
12½	33	5½	23		
13	37	6	27		

Above weights are calculated with Hemp-hearts strands; sol'd made ropes are 1-6th heavier.

CONSECUTIVELY-NUMBERED
Local and Coupon Tickets,
Conductors' Ticket Books and Checks,
CERTIFICATES OF STOCK,
BLANK BOOKS,
ORDERS,
BONDS,
CHECKS,
DRAFTS,
NOTES,
An IMPROVED DATER for LOCAL TICKETS.
ALLEN, LANE & SCOTT,
RAILROAD PRINTING HOUSE
233
SOUTH FIFTH ST., PHILADELPHIA.
RULING AND BOOKBINDING.
MAP BILLS,
MORTGAGES,
TIME TABLES,
MANIFESTS,
WAY-BILLS,
VOUCHERS,
VISITING CARDS,
ENVELOPES, &c.
DISPLAY CARDS and BILLS
PLAIN OR IN COLORS.
DATING MACHINES AND HAND STAMPS.
LOCAL AND COUPON TICKET CASES.
Agents for Engels' Conductor's Punch.

Electric Signals.

Charles T. Chester	New York	See Page 212
The Electric Railroad Signal Co	"	" 184
Towle & Unger Manufacturing Co	"	" 186

Electric Supplies.

Charles T. Chester	New York	See Page 212
F. L. Pope & Co	New York	" 184
Geo. H. Bliss & Co	Chicago, Ill	" 164
Leclanche Battery Co	New York	" 96
L. G. Tillotson & Co	" "	" 94
M. A. Buell	Cleveland, Ohio	" 46
Merchants' Manufacturing & Construction Co	New York	" 18
Towle & Unger Manufacturing Co	" "	" 186

Elevators.

Speedwell Iron Works	New York	See Page 122

Emery.

Champlin & Rogers	Chicago, Ill	See Page 192
D. Brewer & Co	Philadelphia, Pa	" 246
T. B. Bickerton & Co	" "	" 138

Emery Belts.

D. Brewer & Co	Philadelphia, Pa	See Page 246

Emery Grinders.

Champlin & Rogers	Chicago, Ill	See Page 192

GALVANIZED SHEET IRON.

Weight per Square Foot in ounces; Size by Numbers of Wire Gauge.

Size.	Weight	Size.	Weight	Size.	Weight	Size.	Weight
30	10	26	15	22	21	18	37
29	11	25	16	21	24	17	43
28	12	24	17	20	28	16	48
27	14	23	19	19	33	14	60

SHIP SPIKES.

Number in 100 lbs.

Size in Inches. Side.	Length.	No. in 100 lbs.	Size in Inches. Side.	Length.	No. in 100 lbs.	Size in Inches. Side.	Length.	No. in 100 lbs.
1/4	3	1910	7/16	4	542	9/16	6	221
1/4	3½	1585	7/16	4½	503	9/16	6½	200
1/4	4	1326	7/16	5	461	9/16	7	190
1/4	4½	1223	7/16	5½	423	9/16	7½	180
1/4	5	1025	7/16	6	402	9/16	8	170
5/16	3	1010	7/16	6½	321	9/16	8½	160
5/16	3½	963	1/2	5	340	9/16	9	150
5/16	4	810	1/2	5½	312	9/16	10	140
5/16	4½	605	1/2	6	298	5/8	8	140
5/16	5	583	1/2	6½	280	5/8	9	120
5/16	6	521	1/2	7	261	5/8	10	110
			1/2	7½	240	5/8	11	100
			1/2	8	223	3/4	10	80
						3/4	15	60

RAILROAD SPIKES.

Size, Length, Inches.	Thick, Inch.	No. in 100 lbs.	Size, Length, Inches.	Thick, Inch.	No. in 100 lbs.
4½	7-16	351	5½	½	237
4½	½	267	5½	9-16	193
5	⅜	473	5½	⅝	146
5	7-16	326	6	½	207
5	½	260	6	9-16	175
5	9-16	197	6	⅝	131
5	⅝	172			

Emery Wheels.

Champlin & Rogers	Chicago, Ill.	See Page 192
D. Brewer & Co	Philadelphia, Pa	" 246

Engine Builders—Steam.

Buffalo Machinery Agency	Buffalo, N. Y	See Page 160
C. Henry Hall & Co	New York	" 4
Erie City Iron Works	Erie, Pa	" 170
J. B. Waring	New York	" 24
J. S. Mundy	Newark, N. J	" 266

See also, Steam Engines.

Engine Lathes.

Isaac H. Shearman	Philadelphia, Pa	See Page 80
Phelps & Sanger	Cleveland, Ohio	" 120

Engine Trimmings.

J. J. Walworth	Chicago, Ill	See Page 1[illegible]4
J. G. Knapp Manufacturing Co	New York	" 16

Engineers.

Phillipsburg Manufacturing Co	New York	See Page 22

Engineers' Instruments.

F. Eckel	New York	See Page 76
Kuebler & Seelhorst	Philadelphia, Pa	" 216
Wm. J. Young & Sons	" "	" 138

Engineers Stationery Materials.

Charles F. Ketcham	New York	See Page 262

Engines.

The Atlas Works	Pittsburgh, Pa	See Page 172

Engravers.

Chas. F. Ketcham	New York	See Page 262
Mathieson & Schrader	New York	" 82
Warren, Johnson & Co	Buffalo, N. Y	" 160

SPECIFIC GRAVITY AND WEIGHT OF METALS AND ALLOYS.

	Specific Gravity.	Weight of a Cubic Foot. Lbs.	Weight of a Cubic Inch. Lbs.
Aluminium, cast.;	2560	160	.092
Antimony, cast	6710	419	.243
Arsenic	5760	.....	.208
Bismuth, cast	9822	613	.353
Brass, cast {	7800	500	.280
	8400	527	.304
Brasss, Rolled ... about	83[illegible]0	525	.302
Bronze (Copper 8, Tin 1) Gun Metal, about.	8460	528	.305
Copper, cast	8750	543	.315
" , rolled	8800	548	.317
" , wire	8850	553	.321
Gold, pure, cast, 24 carat	19258	1204	.6[illegible]6
" " hammered, 24 carat ... about.	19500	1217	.702
Iron, cast, {purest is} common average	7210	450	.26[illegible]
" wrought {heaviest} " "	7770	485	.280
Lead, cast	113[illegible]0	708	.408
" rolled	11410	712	.411
Mercury— 40°	15630	978	.566
" 32°	13610	848	.492
" 60°	13580	846	.490
" 212°	13375	836	.483
Nickel, Plate	8800	5[illegible]9	.318
Platinium, hammered ... about.	21500	1343	.775
" native, in grains ... " .	17500	.....	.....
Silver, pure, cast	10474	654	.378
" " hammered	10510	655	.379
Steel (Carbon lightens it) common average.	7850	490	.283
Tin (common average)	7291	455	.262
Zinc " "	7000	437	.252

L. G. TILLOTSON & CO.

MANUFACTURERS, IMPORTERS and DEALERS IN

RAILWAY

AND

TELEGRAPH SUPPLIES.

NO. 8 DEY STREET, NEW YORK.

SPECIAL AGENTS AND REPRESENTATIVES OF

The Combination Car Spring Company.
New Haven Car Company.
New Haven Car Head Lining Works.
Buffalo Steam Gauge Company.
Mansfield Elastic Frog Company.
Mercer Rubber Company.
Hamilton Rubber Works.
Pickering's Pulley Blocks.

MANUFACTURERS OF ALL KINDS OF

RAILWAY TRIMMINGS

AND

TELEGRAPH SUPPLIES.

KEEP IN STORE FULL LINE OF ALL MATERIAL FOR

RAILWAY CONSUMPTION.

Exhaust Nozzles.

Geo. W. Richardson & Co............Troy, N. Y............See Page 50

Extra Irons.

J. T. Ryerson............Chicago, Ill............See Page 180

Factory Bells.

A. Fulton's Son & Co............Pittsburgh, Pa............See Page 116

Feather Brushes.

Vose, Dinsmore & Co............New York............See Page 60

Fellys.

Wm. Green & Co............Wilmington, Del............See Page 248

Felt.

H. W. Johns............New York............See Page 224
Vose, Dinsmore & Co............ " " .... " 60

Felting.

Crane Bros. Manufacturing Co............Chicago, Ill............See Page 230

Files.

England & Bindley............Pittsburgh, Pa............See Page 162
G. & H. Barnett............Philadelphia, Pa............ " 216
Geo. Worthington & Co............Cleveland, Ohio............ " 100
Hermann Boker & Co............New York............See Pages 152, 154
Pratt & Co............Buffalo, N. Y............See Page 132
T. B. Bickerton & Co............Philadelphia, Pa............ " 138
Tyng & Co............New York............ " 260

SPECIFIC GRAVITY AND WEIGHT OF WOOD.

	Specific Gravity.	Weight of a Cubic Foot.	Weight of 1000 Feet Board Measure.
		Lbs.	Tons.
Apple	793	49.56	
Ash	752	47.	1.748
" , American White	610	38.	1.414
Beech	690	43.	
Birch	567	35.	
Boxwood	960	60.	
Cherry	672	42.	1.562
Chesnut	550	41.	1.525
Cork	250	15.6	
Dogwood	756	47.	
Ebony, Indian	1209	75.5	
Elm	560	35.	1.302
Fir	512	32.	
Hazel	860	53.7	
Hawthorn	910	56.8	
Hemlock	400	25.	.930
Hickory	850	53.	1.971
Lignum-Vitæ	1333	83.4	
Locust	728	45.5	
Logwood	913	57.	
Mahogany, Spanish	852	53.2	
" , Honduras	560	35.	
Maple	750	46.8	
" , Birds' Eye	576	36.	
Oak, English	777	48.	
" , White	780	49.	
" , Live	950	59.	
" , Red	850	53.	
Pine, White ......average	400	25.	.930
" , Yellow, North, "	550	34.3	1.279
" , " South, "	720	45.	1.674
Poplar	5[illegible]7	36.6	
Satinwood	885	55.3	
Spruce	400	25.	.930
Sycamore	590	37.	1.376
Walnut, Black	610	38.	1.414
Willow	585	36.5	

Fire Brick.

Hall & Sons .... Buffalo, N. Y .... See Page 254
Maurer & Weber .... New York .... " 228
Philip Neukumet .... Philadelphia, Pa .... " 104
Rogers & Co .... Chicago, Ill .... " 144
Star Fire Brick Co .... Pittsburgh, Pa .... " 232
Van Tuyl Manufacturing Co .... New York .... " 242

Fire Buckets.

Darrow, Turner & Co .... New York .... See Page 62

Fire Clay.

Rogers & Co .... Chicago, Ill .... See Page 144

Fire Engines.

Charles B. Hardick .... Brooklyn, N. Y .... See Page 84

Fish Bars.

Hermann Boker & Co .... New York .... See Pages 152, 154
Jones & Laughlins .... Pittsburgh, Pa .... See Page 32

Fish Plates.

Collier & Scranton ; Oxford Iron Co .... New York .... See Page 122
G .B. Walbridge .... " " .... " 130
Greene & Randolph .... " " .... " 54
Lock Nut & Bolt Co. of New York .... " " .... " 272
O. W. Child .... " " .... " 262
Tyng & Co .... " " .... " 260

Flags.

De Grauw, Aymar & Co .... New York .... See Page 112

Flange Pipes.

R. A. Brick & Co .... New York .... See Page 48

SHAFTS AND WELLS.

(Condensed from Trautwine.)

	For Shafts.	For Wells.			For Shafts.	For Wells.	
Diam. in Feet.	Cubic yds. of Digging.	No. Bricks in lining 1 br'k thick.	Conts. in Gals.	Diam. in Feet.	Cubic yds. of Digging.	No. Bricks in lining 1 br'k thick.	Conts. in Gals.
3	.262	170	.367	14	5.7	792	7.997
½	.356	198	.499	½	6.116	820	8.578
4	.465	227	.653	15	6.545	849	9.18
½	.589	255	.826	½	6.99	877	9.8
5	.727	283	1.02	16	7.447	905	10.44
½	.880	311	1.234	½	7.92	933	11.11
6	1.047	340	1.47	17	8.407	962	11.8
½	1.229	368	1.724	½	8.908	990	12.5
7	1.425	396	2.	18	9.425	1018	13.22
½	1.636	425	2.295	½	9.95	1046	13.96
8	1.862	453	2.611	19	10.5	1075	14.73
½	2.102	481	2.948	½	11.06	1103	15.51
9	2.356	509	3.305	20	11.64	1131	16.32
½	2 625	538	3.682	½	12.22	1160	17.15
10	2.91	566	4.08	21	12.83	1188	18.
½	3.207	594	4.498	½	13.45	1216	18.85
11	3.52	622	4.937	22	14.08	1244	19.75
½	3.847	651	5.396	½	14.73	1273	20.66
12	4.19	679	5.875	23	15.39	1301	21.58
½	4.54	707	6.375	½	16.06	1329	22.53
13	4.916	736	6.895	24	16.76	1357	23.5
½	5.3	764	7.436	25	18.18	1414	25.5

For Shafts, or for cubic yards of digging for Wells, take the diameter of the digging; for diameters greater than given, take a given size ½ the required one, and multiply the quantity by 4; thus, for 40 feet take 11.64 (for 20 feet), multiply by 4 = 46.56 cubic yards.

For Wells, take the diameter in clear of lining, and for bricks in larger sizes take double the quantity given above for ½ the required size; the brick lining will be 8¼ inches thick; for gallons, take diameter inside of lining, and for larger diameters, observe same rule as given above for cubic yards of digging.

DeLaney & Co.

BUFFALO, NEW YORK.

Niagara Forge Works.

FINISHED and ROUGH, MARINE and STATIONARY ENGINE, RAILROAD ENGINE, and CAR FORGINGS,

HAMMERED SCRAP

A SPECIALTY

Iron first prepared under a TEN-TON hammer.

AXLES CENTERED READY FOR LATHES,

AND

DATED MONTH AND YEAR OF MANUFACTURE.

Cheapest and Best Hammered Axle Sold.

☞ References cheerfully given.

(*ESTABLISHED IN 1850.*)

W. P. KELLOGG & CO.

TROY, N. Y., and 118 Chambers St., NEW YORK.

"The Empire"

A FAN BLOWING PORTABLE FORGE,

FOR RAILROADS, QUARRIES, MINES, MACHINE AND BOILER SHOPS,

Is the best, most durable, simplest, cheapest and most effective forge ever made.

ALSO

Boring Machines, Mortising Machines,

SCALES, CURRY COMBS and HARDWARE SPECIALTIES.

☞ SEND FOR A CIRCULAR.

Flue Iron.

Wm. Green & Co.........Wilmington, Del..........See Page 248

Fog Signals—Torpedo.

Tyng & Co.........New York..........See Page 260

Foot Lathes.

Phelps & Sanger.........Cleveland, Ohio..........See Page 120
See also Lathes.

Force Pumps.

A. Carr.........New York..........See Page 244
C. Henry Hall & Co......... " " .... " 4
Vose, Dinsmore & Co......... " " .... " 60

Forge Bellows.

Geo. Worthington & Co.........Cleveland, Ohio..........See Page 100
Joseph Churchyard.........Buffalo, N. Y.......... " 160
Newcomb Bro's. Sons.........New York.......... " 88

Forges.

John Bayliss.........New York..........See Page 76
Newcomb Bro's. Sons......... " " .... " 88
S. S. Townsend......... " " .... " 2
Wm. B. Bement & Son.........Philadelphia, Pa.......... " 168
W. P. Kellogg & Co.........Troy, N. Y.......... " 98

VALUE OF IRON.

Value per Gross Ton (2240 lbs.) of IRON, at from 1-16th of a cent to 10 cents per Pound, increasing at Rate of 1-16th and 1-8th of a cent per Pound.

Per lb. in cts. & 1-16ths.	Price per Ton.	Per lb. in cts. & 1-16ths.	Price Per Ton.	Per lb. in cts. & 1-8ths.	Price per Ton.
1-16	$ 1.40	2 5-8	$58.80	7-8	$165.20
1-8	2.80	11-16	60.20	1-2	168.
3-16	4.20	3-4	61.60	5-8	170.80
1-4	5.60	13-16	63.	3-4	173.60
5-16	7.	7-8	64.40	7-8	176.40
3-8	8.40	15-16	65.80	8	179.20
7-16	9.80	3	67.20	1-8	182.
1-2	11.20	1-8	70.	1-4	184.80
9-16	12.60	1-4	72.80	3-8	187.60
5-8	14	3-8	75.60	1-2	190.40
11-16	15.40	1-2	78.40	5-8	19[illegible].20
3-4	16.80	5-8	81.20	3-4	196.
13-16	18.20	3-4	84.	7-8	198.80
7-8	19.60	7-8	86.80	9	201.60
15-16	21.	4	8[illegible].60	1-8	204.40
1	22.40	1-8	92.40	1-4	207.20
1-16	23.80	1-4	95.20	3-8	210.
1-8	25.20	3-8	98.	1-2	2[illegible]2.80
3-16	26.60	1-2	100 80	3-4	218,40
1-4	28.	5-8	103.60	10	224.
5-16	29.40	3-4	106. 0	1-8	226 80
3-8	30.80	7-8	109.20	1-4	229.60
7-16	32.20	5	112.	3-8	232.40
1-2	33.60	1-8	114.80	1-2	23[illegible].20
9-16	35.	1-4	117.60	5-8	238.
5-8	36.40	3-8	120.40	3-4	240.80
11-16	37.80	1-2	123.20	7-8	243.60
3-4	39.20	5-8	126.	11	246.40
13-16	40.60	3-4	128.80	1-8	249 20
7-8	42.	7-8	131.60	1-4	252
15-16	43.40	6	134.40	3-8	254.80
2	44.80	1-8	137.20	1-2	257.60
1-16	46.20	1-4	140.	5-8	260 40
1-8	47.60	5-8	142.80	3-4	263.20
3-16	49.	1-2	145.60	7-8	266.
1-4	50.40	5-8	148.40	12	268.80
5-16	51.80	3-4	151.20	1-8	271 60
3-8	53.20	7-8	154.	1-4	274.40
7-16	54.60	7	156.80	3-8	277 20
1-2	56.	1-8	159.60	1-2	280.
9-16	57.40	1-4	162.40		

Forgings.

DeLaney & Co. .... Buffalo, N. Y. .... See Page 98
Pittsburgh Forge & Iron Co .... Pittsburgh, Pa .... " 174

Founders.

Snyder Bros .... Williamsport, Pa .... See Page 46

Founders' Tools.

S. S. Townsend .... New York .... See Page 2

Foundry Facings.

Van Tuyl Manufacturing Company .... New York .... See Page 242

Foundry Fixtures.

Wm. B. Bement & Son .... Philadelphia, Pa .... See Page 168

Foundry Supp ies.

Van Tuyl Manufacturing Co .... New York .... See Page 242

Freight Cars.

Chas. W. Matthews .... Philadelphia, Pa .... See Page 38
Greene & Randolph .... New York .... " 54
O. W. Child .... " " .... " 262

See also, Car Builders.

Frogs.

Anderson & Woods .... Pittsburgh, Pa .... See Page 150
L. G. Tillotson & Co .... New York .... " 94
O. W. Child .... " " .... " 262
Pittsburgh Bolt Co .... Pittsburgh, Pa .... " 118
Skinner & Gifford Manufacturing Co .... Dunkirk, N. Y .... " 140
The Atlas Works .... Pittsburgh, Pa .... " 172

VALUE OF IRON.

Value per Gross Ton (2240 lbs.) of IRON at from 1-10th of a cent to 10 cents per Pound, increasing at Rate of 1-10th of a cent per Pound.

Per lb. in cts. & 1-10ths.	Price per Ton.	Per lb. in cts. & 1-10ths.	Price per Ton.	Per lb. in cts. & 1-10ths.	Price per Ton.
1-10	$ 2.24	3 5-10	$ 78.40	6 8-10	$152.32
2-	4.48	6-	80.64	9-	154.56
3-	6.72	7-	82 88	7	156.80
4-	8.96	8-	85.12	1-10	158.04
5-	11.20	9-	87.36	2-	161.28
6-	13.44	4	89.60	3-	163.52
7-	15.68	1-10	91.84	4-	165.76
8-	17.92	2-	94.08	5-	168.
9-	20 16	3-	96.32	6-	170.24
1	22.40	4-	98 56	7-	172.48
1-10	24.64	5-	100.80	8-	174.72
2-	26.88	6-	103 04	9-	176.96
3-	29.12	7-	105 28	8	179.20
4-	31.36	8-	107.52	1-10	181.44
5-	33.60	9-	109.76	2-	183.68
6-	35 84	5	112.00	3-	185.92
7-	38.08	1-10	114.24	4-	188.16
8-	40.32	2-	116.48	5-	190.40
9-	42.56	3-	118.62	6-	192.64
2	44.80	4-	120.96	7-	194.88
1-10	47.04	5-	123.20	8-	197.12
2-	49 28	6-	125.44	9-	199.36
3-	51.52	7-	127.68	9	201.60
4-	53.76	8-	129 92	1-10	203.84
5-	56.	9-	132.16	2-	206 08
6-	58.24	6	134.40	3-	208.32
7-	60.48	1-10	136.64	4-	210.56
8-	62.72	2-	138.88	5-	212 80
9-	64.96	3-	141.12	6-	215 04
3	67.20	4-	143.36	7-	217.28
1-10	69.44	5-	145.60	8-	219.52
2-	71.68	6-	147.84	9-	221.76
3-	73.92	7-	150.08	10	224.
4-	76.16				

E. A. C. DU PLAINE. PAUL S. REEVES.

TUBAL SMELTING WORKS,

No. 760 So. Broad St., PHILADELPHIA.

DU PLAINE & REEVES,

Brass Founders and Smelters

Anti-Friction or Babbitt Metals.

XXX METAL, Nick. Hardening, 55c.
XX " " " 50c.
X " Copper " 45c.

Guaranteed to contain no lead.

A METAL. 40c. **B METAL.** 35c.
C " 30c. **D "** 25c.
E METAL. 20c.

These metals contain a very large per centage of tin, antimony, and copper hardening, according to price.

F METAL. 18c. **G METAL.** 16c
H " 14c. **J "** 12c.

These metals are the ordinary low-priced Babbitt Alloys, used where there is not much wear on the machinery, and where economy is required. They are *superior* to the higher-priced metals of other markets.

Brass Castings.

ENGINE BRASS, ROLLING MILL BRASS, PACKING RINGS, &c.
First quality metals, warranted new, (copper and tin.)

ORDINARY RED BRASS CASTINGS,
of superior metals, (part new, part old.)

GOOD QUALITY RED BRASS CASTINGS,
from well selected old metal.

No. 1, Yellow Brass Castings,
all new metal.
" 2, Yellow Brass Castings.
" 1, Car-Journal Boxes,
new metal.
" 2, Car-Journal Boxes,
one-half new, one-half old metal.
" 3, Car-Journal Boxes,
good old metal.
" 4, Car-Journal Boxes,
common metal.

Ingot, or Pig Brass

We manufacture this material for Brass Founders and Manufacturers, who prefer to do their own Casting; and will supply them at lower rates than they can purchase elsewhere. A written guarantee given with all purchases, if desired, of new metal, subject to analysis at our cost, if not as represented.

10 Parts Copper to 1 Tin. **8 Parts Copper, to 1 Tin,** } New
9 " " 1 " **7 " " 1 "** }
6 Parts Copper to 1 Tin, } Metal.

The same mixtures from well-selected old metal, with part new copper and tin added to restore it.

Ordinary Red Brass, no new metal in pig.
No. 1. Yellow Brass, new metal, 1½ to 4 parts of Copper to 1 of Spelter, in pig or ingots, warranted.
No. 2. Yellow Brass, old and new mixed.
No. 3. All Old Yellow Brass, mixed.

☞ *SEND FOR PRICE LIST.*

CONTENTS OF CYLINDERS AND PIPES.

In Cubic Feet and Gallons.

(Condensed from Trautwine.)

Diameter in Ins.	For 1 Foot in length. Cubic Feet.	For 1 Foot in length. Gallons of 231 Cub. Ins.	Diameter in Ins.	For 1 Foot in length. Cubic Feet.	For 1 Foot in length. Gallons of 231 Cub. Ins.	Diameter in Ins.	For 1 Foot in length. Cubic Feet.	For 1 Foot in length. Gallons of 231 Cub. Ins.
¼	.0003	.0025	4	.0873	.6528	10	.5454	4.08
5/16	.0005	.004	¼	.0985	.7369	¼	.573	4.286
⅜	.0008	.0057	½	.1104	.8263	½	.6013	4.498
7/16	.001	.0078	¾	.1231	.9206	¾	.6303	4.715
½	.0014	.0102	5	.1364	1.02	11	.66	4.937
9/16	.0017	.0129	¼	.1503	1.125	¼	.6903	5.164
⅝	.0021	.0159	½	.165	1.234	½	.7213	5.396
11/16	.0026	.0193	¾	.1803	1.349	¾	.753	5.633
¾	.0031	.0230	6	.1963	1.469	12	.7854	5.875
13/16	.0036	.0269	¼	.2131	1.594	½	.8522	6.375
⅞	.0042	.0312	½	.2304	1.724	13	.9218	6.895
15/16	.0048	.0359	¾	.2485	1.859	½	.994	7.436
1	.0055	.0408	7	.2673	1.999	14	1.069	7.997
¼	.0085	.0638	¼	.2867	2.145	½	1.147	8.578
½	.0123	.0918	½	.3068	2.295	15	1.227	9.180
¾	.0167	.1249	¾	.3276	2.45	½	1.31	9.801
2	.0218	.1632	8	.3491	2.611	16	1.396	10.44
¼	.0276	.2066	¼	.3712	2.777	½	1.485	11.11
½	.0341	.255	½	.3941	2.948	17	1.576	11.79
¾	.0412	.3085	¾	.4176	3.125	½	1.67	12.49
3	.0491	.3672	9	.4418	3.305	18	1.767	13.22
¼	.0576	.4309	¼	.4667	3.491	½	1.867	13.96
½	.0668	.4998	½	.4922	3.682	19	1.969	14.73
¾	.0767	.5738	¾	.5185	3.879	½	2.074	15.51
						20	2.182	16.32

To find the contents of a larger pipe than given above, take one-half the size and multiply by 4, or take one-fourth the size and multiply by 16. Thus: wanted, the contents of a pipe 30 inches in diameter; 9.180 (contents of 15 inch pipe) × 4 = 36.72 gallons. Wanted, contents of a pipe 50 inches in diameter; .8522 (contents of 12½ inch pipe) × 16 = 13.6352 cubic inches.

Galvanized Pipe.

Evans, Dalzell & Co....................Pittsburgh, Pa..............See Page 58
Iorris, Tasker & Co....................Ph ladelphia, Pa............ " 36

Galvanizers.

ohn Merry & Co....................New York....................See Page 48
Iarshall Lefferts, Jr.................... " " .... " 44

Galvanometers.

harles T. Chester....................New York....................See Page 212
Ho. H. B iss & Co....................Chicago, Ill .... " 164
eclanche Battery Co....................New York.................... " 96

Gas Apparatus.

Iorris, Tasker & Co....................Philadelphia, Pa............See Page 36

Gas and Steam Fittings.

. Fulton's Son & Co....................Pittsburgh, Pa..............See Page 116
rane Bros. Manufacturing Co....................Chicago, Ill.................... " 230
IcNab & Harlin Manufacturing Co....................New York.................... " 220
Iorris, Tasker & Co....................Philadelphia, Pa.................... " 36
ancoast & Maule.................... " " .... " 26

FALLING BODIES.

Velocity Due to Different Heights (Approximate.)

(Molesworth.)

Fall in feet.	Velocity. Feet per second.	Fall in feet.	Velocity. Feet per second.	Fall in feet.	Velocity. Feet per second.
1	8	50	57	275	133
2	11.3	60	62	300	139
3	13.9	70	67	325	144
4	16	80	72	350	150
5	18	90	76	375	155
10	25	100	80	400	160
15	31	125	90	450	170
20	36	150	98	500	179
25	40	175	106	550	188
30	44	200	113	600	196
35	47	225	120	800	227
40	51	250	127	1000	254

RIVETED COPPER PIPES.

Weight of One Running Foot.

Including the necessary laps for riveting or calking, but not including weight of rivets, which will, of course, vary according to number and sizes used, in proportion to thickness of metal.

Diam. Inch.	Thickness, in Inches. 1/8	3/16	1/4	5/16
5	8.16	12.25	16.30	
½	8.9	13.35	17.8	
6	9.65	14.5	19.3	
½	10.5	15.7	20.9	
7	11.2	16.8	22.4	
½	12.	18.	24.	
8	12.6	19.1	25.5	
½	13.5	20.2	27.	
9	14.3	21.5	28.6	
½	15.	22.5	30.1	
10	16.	24.	32.	40.
½	—	25.2	33.5	41.8
11	—	26.3	35.	43.7
12	—	28.5	38.	47.5
13	—	31.2	41.5	51.8
14	—	33.2	44.3	55.4

Diam. Inch.	Thickness, in Inches. 3/16	1/4	5/16
15	35.5	47.2	59.3
16	37.8	50.5	62.9
17	39.8	53.2	66.5
18	42.4	56.5	70.5
19	44.7	59.5	75.
20	—	62.6	78.
21	—	66.	82.5
22	—	69.2	86.5
23	—	72.3	90.2
24	—	75.5	93.6
25	—	78.7	97.5
26	—	80.2	100.5
27	—	83.6	104.5
28	—	86.8	108.5
29	—	90.2	112.7
30	—	92.4	116.

Gas and Steam Fitters' Tools.

Champlin & Rogers	Chicago, Ill	See Page 192
Crane Bros. Manufacturing Co	" "	" 230
McNab & Harlin Manufacturing Co	New York	" 220
Morris, Tasker & Co	Philadelphia, Pa	" 36
Nelson Tool Works	New York	" 30
Pancoast & Maule	Philadelphia, Pa	" 26
Phelps & Sanger	Cleveland, Ohio	" 120

Gas Pipe.

A. Carr	New York	See Page 244
Evans, Dalzell & Co	Pittsburgh, Pa	" 58
J. J. Walworth	Chicago, Ill	" 124
J. T. Ryerson	" "	" 180
Morris, Tasker & Co	Philadelphia, Pa	" 36
Phelps & Sanger	Cleveland, Ohio	" 120
R. A. Brick & Co	New York	" 48
Winans & Co	" "	" 106
Wm. Graff & Co	Pittsburgh, Pa	" 174

Gauges.

Buffalo Steam Gauge Co	Buffalo, N. Y	See Page 170
J. G. Knapp Manufacturing Co	New York	" 16
McNab & Harlin Manufacturing Co	" "	" 220

Gauge Tubes.

J. J. Walworth	Chicago, Ill	See Page 124

Geared Foot Lathes.

Wm. Sellers & Co	Philadelphia, Pa	See Page 64

FLAT IRON.

Number of Feet in a Bundle of 112 Pounds.

Size.	Feet in Bundle.	Size.	Feet in Bundle.
1-2 by 1-4 Inch.	267	7-8 by 1-4 Inch.	155
1-2 " 5-16 "	216	7-8 " 5-16 "	122
1-2 " 3-8 "	175	7-8 " 3-8 "	100
5-8 " 1-4 "	214	7-8 " 7-16 "	90
5-8 " 5-16 "	170	7-8 " 1-2 "	75
5-8 " 3-8 "	145	7-8 " 5-8 "	60
5-8 " 1-2 "	106	1 " 1-4 "	135
3-4 " 1-4 "	175	1 " 5-16 "	106
3-4 " 5-16 "	142	1 " 3-8 "	90
3-4 " 3-8 "	120	1 " 7-16 "	78
3-4 " 7-16 "	103	1 " 1-2 "	65
3-4 " 1-2 "	90	1 " 9-16 "	60
3-4 " 5-8 "	70	1 " 5-8 "	52

ROUND AND SQUARE IRON.

Number of Feet in a Bundle of 112 Pounds.

Round Iron.		Square Iron.	
Size.	Feet in Bundle.	Size.	Feet in Bundle.
3-16 Inch.	1 15	3-16 Inch.	958
1-4 "	688	1-4 "	540
5-16 "	440	5-16 "	345
3-8 "	305	3-8 "	240
7-16 "	225	7-16 "	176
1-2 "	170	1-2 "	135
9-16 "	136	9-16 "	107
5-8 "	110	5-8 "	87
11-16 "	90	11-16 "	70
3-4 "	75	3-4 "	60

PULLEYS OR BLOCKS.

A movable pulley reduces the necessary amount of power, in the proportion ot 2 to 1 for every wheel in the pulley; thus, a movable pulley with 1 wheel will reduce the power one-half, 2 wheels to one-fourth, 3 wheels to one-sixth, etc.

Girders.

Rogers & Co. .... Chicago, Ill. .... See Page 144

Glass Bearings.

John Harden .... New York .... See Page 230

Glass Letters.

Jay & Cook .... New York .... See Page 234

Glass—Looking.

D. R. Hobart & Co. .... New York .... See Page 126
Jay & Cook .... " " .... " 234
Leffingwell & Co. .... Cleveland, Ohio .... " 190

Glass—Ornamented.

D. R. Hobart & Co. .... New York .... See Page 126
Jay & Cook .... " " .... " 234
Leffingwell & Co. .... Cleveland, Ohio .... " 190

Glass—Plate.

D. R. Hobart & Co. .... New York .... See Page 126
Jay & Cook .... " " .... " 234
Leffingwell & Co. .... Cleveland, Ohio .... " 190

Glass—Window.

D. R. Hobart & Co. .... New York .... See Page 126
Jay & Cook .... " " .... " 234
Leffingwell & Co. .... Cleveland, Ohio .... " 190

Glaziers' Points.

D. R. Hobart & Co. .... New York .... See Page 126
G. B. Walbridge .... " " .... " 130
Leffingwell & Co. .... Cleveland, Ohio .... " 190
Jay & Cook .... New York .... " 234

Globe and Angle Valves.

A. Fulton's Son & Co. .... Pittsburgh, Pa. .... See Page 116

Gongs.

J. J. Walworth .... Chicago, Ill. .... See Page 124
Radley & McAlister Manufacturing Co. .... New York .... " 250

Grindstones.

W. R. Wood & Co. .... New York .... See Page 210

BAND IRON.

Number of Feet in a Bundle of 112 Pounds.

Size, Width.		Thick.	Feet in Bundle.	Size, Width.		Thick.	Feet in Bundle.
1⅛	Inches	No. 12	265	2¾	Inches	No. 12	110
1⅛	"	" 10	213	2¾	"	" 10	88
1⅛	"	" 7	160	2¾	"	" 8	72
1¼	"	" 12	246	2¾	"	" 6	60
1¼	"	" 10	190	3	"	" 12	101
1¼	"	" 7	145	3	"	" 10	80
1½	"	" 12	205	3	"	" 8	66
1½	"	" 10	160	3	"	" 6	57
1½	"	" 7	120	3¼	"	" 10	75
1¾	"	" 12	175	3¼	"	" 8	60
1¾	"	" 10	138	3¼	"	" 6	50
1¾	"	" 8	110	3½	"	" 10	69
1¾	"	" 7	100	3½	"	" 8	57
2	"	" 12	155	3½	"	" 6	48
2	"	" 10	120	4	"	" 10	60
2	"	" 8	99	4	"	" 8	50
2	"	" 7	90	4	"	" 6	40
2	"	" 6	81	4½	"	" 10	52
2¼	"	" 12	135	4½	"	" 8	43
2¼	"	" 10	105	4½	"	" 6	35
2¼	"	" 8	88	5	"	" 10	48
2¼	"	" 6	72	5	"	" 8	40
2½	"	" 12	120	5	"	" 6	34
2½	"	" 10	95	6	"	" 10	40
2½	"	" 8	77	6	"	" 8	32
2½	"	" 6	65	6	"	" 6	26

Sixth.

IT WILL SAVE OVER ONE-HALF OF ALL THE REPAIRS TO BOILERS, by keeping all parts of the iron, the water, and the steam at a practically equal temperature. Thus the strains of unequal expansion and contraction are avoided.

Seventh.

It will more than **DOUBLE** the time a boiler will last in a perfectly serviceable condition.

Eighth.

IT WILL SAVE A LARGE PROPORTION OF THE REPAIRS TO THE ENGINE. By insuring dry steam, not surcharged with water, no loose sediment or grit is carried over to cut the valves, piston or cylinder, and no solid water is presented between the piston and cylinder heads to cause pounding, to break or derange the head or flanges of the cylinder or other parts of the engine.

Ninth.

It may save a large proportion of the oil used for lubricating the cylinder and valves of the engine. In fact, all devices for supplying oil to cylinders of high pressure engines may be safely abolished, as the Boiler Attachment will insure the supply of the best quality of steam for lubrication; and steam, neither superheated nor surcharged with water, is the best kind of a lubricator, and is always present at the time and place it is required.

While the

WIARD STEAM BOILER ATTACHMENT

was founded upon the idea or theory that *EXPLOSIONS, RUPTURES, AND LEAKS OF STEAM BOILERS* were caused by unequally heating or cooling the different parts of the boiler, it will absolutely prevent such accidents or injury to the boilers on which it may be applied, even upon the supposition that any other theory ever advanced or advocated is or may be the correct one.

No person, since it was invented, has been ingenious enough to invent or suggest any possible way to injure a boiler on which it has been applied.

While we will undertake to explode or destroy any boiler whatever, even with the safety valves in good order, and without over-pressure on which the attachment is not applied, and with only the ordinary fires, we defy any engineer to injure a boiler in the slightest degree on which it is applied, either by subjecting it to the same conditions or any other, except disarranging the safety valve.

For further information, apply to

The Wiard Locomotive Attachment Co.,

23 DEY ST., NEW YORK.

BRUCE & COOK,

IMPORTERS of METALS

186, 188 & 190 Water Street, and 248 & 250 Pearl Street,

NEW YORK.

TIN PLATE, SHEET IRON, COPPER, ANTIMONY, LEAD, SHEET ZINC, BLOCK TIN, GALVANIZED SHEET IRON, TINNED and BRIGHT WIRE, ANNEALED FENCE WIRE, TINSMITHS' TOOLS and MACHINES, BAR and SHEET LEAD and LEAD PIPE.

JOHN M. BRUCE, JR. JOHN C. COOK. RUSSELL W. McKEE.

PHILIP S. MILLER. *LENOX SMITH.*

AMERICAN and FOREIGN

STEEL AND IRON RAILS

Street Rails, Railway Fastenings, &c.

Locomotives, Cars and Machinery.

The Dickson Manufacturing Company of Scranton.

MILLER & SMITH,

43 EXCHANGE PLACE, - NEW YORK.

DEGRAUW, AYMAR & CO.

IMPORTERS and DEALERS IN

CHAINS, ANCHORS, BUNTING,

RUSSIA BOLT ROPE and WIRE ROPE.

MANUFACTURERS OF

CORDAGE AND OAKUM.

42 SOUTH STREET, NEW YORK.

The Highest Grades of Crane Chains a Specialty.

REFINED TALLOW FOR CYLINDERS.

HEMPEN ROPES AND CABLES—SAFE LOADS.

(*From Haslett.*)

ROPES.

To find the Strain that may be applied to a Hempen Rope with Safety.

Circumference.	Pounds.	Circumference.	Pounds.	Circumference.	Pounds.
1	200	3½	2450	6	7200
¼	312	¾	2812	¼	7812
½	450	4	3200	½	8450
¾	612	¼	3612	¾	9112
2	800	½	4050	7	9800
¼	1012	¾	4512	¼	10512
½	1250	5	5000	½	112[illegible]0
¾	1512	¼	5512	¾	12012
3	1800	½	6050	8	12800
¼	2112	¾	6612		

CABLES.

To find the Strain that may safely be applied to a good Hempen Cable.

Circumference.	Pounds.	Circumference.	Pounds.	Circumference.	Pounds.
6	4320	10¼	12607	14½	25230
¼	4687	½	13230	¾	26107
½	5070	¾	13867	15	27000
¾	5467	11	14520	¼	27907
7	5880	¼	15187	½	28830
¼	6307	½	15870	¾	29767
½	6750	¾	16567	16	30720
¾	72[illegible]7	12	17280	¼	31687
8	7680	¼	18007	½	32670
¼	8167	½	18750	¾	33667
½	8679	¾	19507	17	34680
¾	9187	13	20280	¼	35707
9	9720	¼	21067	½	36750
¼	10267	½	21870	¾	37807
½	10830	¾	22687	18	38880
¾	11407	14	23520	¼	39967
10	12000	¼	24367		

To Ascertain the Strength of Cables.—Multiply the square of the circumference in inches by 120, and the product is the weight the cable will bear in pounds, with safety.

To Ascertain the Strength of Ropes.—Multiply the square of the circumference in inches by 20[illegible], and it gives the weigth the rope will bear in pounds, with safety.

To Ascertain the Weight of Manilla Ropes and Hawsers.—Multiply the square of the circumference in inches by .03, and the product is the weight in pounds of a foot in length.

This is but an approximation, sufficiently correct for many purposes.

NEW YORK LANTERN COMPANY,

MANUFACTURERS OF

Railroad, Ship,

AND

Hand Lanterns,

Woodward's, French's Screw, Excelsior Reflector,

Sweeny's and Avery's Railroad Lanterns

AND

AMBROSE'S PATENT NO CHIMNEY

STREET AND CAR LAMPS.

—— ALSO, ——

Street Lamps for Light Oil with Patent Burners.

S. M. AIKMAN & CO.,

261 PEARL ST., NEW YORK.

Hand Forcing Machines.

Wm. Sellers & Co.....Philadelphia, Pa.....See Page 64

Hand Lamps.

S. M. Aikman & Co.....New York.....See Page 114

Hand Lathes.

Wm. Sellers & Co.....Philadelphia, Pa.....See Page 64

Hand Punches.

Biddle Manufacturing Co.....New York.....See Page 12

Hand Shears.

Biddle Manufacturing Co.....New York.....See Page 12

Hand Stamps.

Mathiesen & Schrader.....New York.....See Page 82

Hangers, &c.

Buffalo Machinery Agency.....Buffalo, N. Y.....See Page 160
Geo. V. Cresson.....Philadelphia, Pa..... " 138
J. S. Mundy.....Newark, N. J..... " 226
R. L. Howard & Son.....Buffalo, N. Y..... " 40
Wm. Sellers & Co.....Philadelphia, Pa..... " 64
See also, Shafting.

Hardware.

Biddle Manufacturing Co.....New York.....See Page 12
D. Brewer & Co.....Philadelphia, Pa..... " 246
England & Bindley.....Pittsburgh, Pa..... " 162
G. B. Walbridge.....New York..... " 130
Geo. Worthington & Co.....Cleveland, Ohio..... " 100
H. B. Newhall, Agent.....New York..... " 166
Hermann Boker & Co..... " " .....See Pages 152, 154
Sidney Shepard & Co.....Buffalo, N. Y.....See Page 196

Hard Wood Doors.

Joseph Churchyard.....Buffalo, N. Y.....See Page 160

Harness.

C. M. Moseman & Bro.....New York.....See Page 226

EFFECT OF HEAT UPON VARIOUS BODIES.

(Condensed from Haswell.)

Wedgewood's zero is 1077° of Fahrenheit, and each degree = 130°.

In the designation of the degrees of temperature, the symbol + is omitted when the temperature is above 0; but when it is below, the symbol − is prefixed.

	Degrees.
Ammonia Boils	140
Ammonia (liquid) freezes	−46
Antimony melts	951
Arsenic melts	365
Bismuth melts	476
Blood (human) heat of	98
" " freezes	25
Brandy freezes	−7
Brass melts	1900
Cadmium melts	600
Coal Tar boils	325
Cold, greatest artificial	−166
" greatest natural	−56
Common Fire	790
Copper melts	2548
Glass melts	2377
Gold (fine) melts	2590
Gutta-percha softens	145
Heat, cherry red	1500
" " (Daniell)	1141
" bright red	1860
" red visible by day	1077
" white	2900
Ice melts	32
Iron (cast) melts	3479
" (wrought) melts	3980
Iron, bright red in the dark	752
" red hot in twilight	884
Lead melts	504
Mercury boils	662
" volatilizes	680
" melts	−39
Naptha boils	186
Petroleum boils	306
Platinum melts	3080
Potassium melts	135
Proof Spirit freezes	−7
Saltpetre melts	600
Sea-water freezes	28
Silver (fine) melts	1250
Snow and Salt, equal parts	0
Spirits of Turpentine freezes	14
Steel melts	2500
" polished, blue	580
" " straw color	460
Strong Wines freeze	20
Sulphur melts	226
Sulph. Acid (sp. grav. 1.641) freezes	−45
Tin melts	421
Vinous fermentation 60 to	77
Water, in *vacuo* boils	98
Zinc melts	740

Fulton's Bell AND Brass Foundry,

91 First Avenue & 70 Second Avenue,

PITTSBURGH, PA.

A. FULTON'S SON & CO.

MANUFACTURERS OF

BELLS

FOR

CHURCHES,
SCHOOLS,
LOCOMOTIVES,
FACTORIES,
STEAMBOATS,
ENGINES,

AND

ALL OTHER PURPOSES.

ALSO,

Car Bearings, Brass Castings,
Babbitt Metal, Steam Gauges,
Globe and Angle Valves,

AND DEALERS IN

Railroad Supplies.

☞ SEND FOR CATALOGUE. ☜

Harrow Teeth.

Hermann Boker & Co........ ....New York...........See Pages 152, 154

Hasps.

D. Brewer & Co..Philadelphia, Pa............See Page 246

Hassocks—Carpet.

New York Ottoman & Hassock Co.................New York.....................See Page 236
Geo. Drake Smith...Philadelphia, Pa........... " 246

Head Lamps and Lights.

L. G. Tillotson & Co......................................New York............... . ...See Page 94
Radley & McAlister Manufacturing Co........... " " .... " 250
Vose, Dinsmore & Co.................................. " " .... " 60
Wm. B. Browne & Co.................................. " " .... " 266

Head Light Oil.

F. S. Pease..Buffalo, N. Y..................See Page 100
Wm. B. Browne & Co..................................New York...................... " 266

BAR AND SHEET COPPER.

Weight in Pounds.

Thickness, or Diameter, or Side; Inches.	Sheets per Square Foot.	Square Bars 1 Foot Long.	Round Bars 1 Foot Long.	Thickness, or Diameter, or Side; Inches.	Sheets per Square Foot.	Square Bars 1 Foot Long.	Round Bars 1 Foot Long.
1/16	2.88	.015	.011	1 1/16	49.	4.35	3.41
1/8	5.75	.06	.046	1/8	52.	4.86	3.85
3/16	8.65	.134	.105	3/16	54.9	5.40	4.29
1/4	11.48	.235	.187	1/4	57.65	6.	4.73
5/16	14.36	.375	.295	5/16	60.5	6.6	5.20
3/8	17.28	.54	.424	3/8	63.45	7.27	5.70
7/16	20.19	.735	.575	7/16	66.35	7.90	6.28
1/2	23.1	.960	.75	1/2	69.3	8.64	6.80
9/16	26.	1.21	.95	9/16	72.15	9.28	7.30
5/8	28.85	1.51	1.17	5/8	75.1	10 15	8.
11/16	31.68	1.81	1.42	11/16	77.95	10.95	8.6
3/4	34.57	2.15	1.7	3/4	80.75	11.70	9.24
13/16	37.46	2.54	2.	13/16	83.60	12.60	9.85
7/8	40.39	2.95	2.3	7/8	86.58	13.46	10.55
15/16	43.27	3.37	2.64	15/16	89.45	14.35	11.25
1	46.15	3.84	3.01	2	92.25	15.35	12.

BAR AND SHEET LEAD.

Weight in Pounds.

Thickness, or Diameter, or Side; Inches.	Sheets per Square Foot.	Square Bars 1 Foot Long.	Round Bars 1 Foot Long.	Thickness or Diameter, or Side; Inches.	Sheets per Square Foot.	Square Bars 1 Foot Long.	Round Bars 1 Foot Long.
1/16	3.71	.02	.014	1 1/16	63.2	5.6	4.4
1/8	7.43	.079	.06	1/8	66.87	6.26	4.91
3/16	11.	.175	.136	3/16	70 51	6.98	5.5
1/4	14.08	.31	.245	1/4	74.35	7.74	6.1
5/16	18.05	.486	.38	5/16	78.05	8.55	6.73
3/8	22.02	.695	.549	3/8	81.76	9.38	7.38
7/16	26.	.948	.745	7/16	85.48	10.18	8.05
1/2	29.75	1.25	.975	1/2	89.28	11.0	8.75
9/16	33.49	1.55	1.24	9/16	93.	12.05	9.50
5/8	37.18	1.95	1.51	5/8	96.78	13.15	10.25
11/16	40.87	2.33	1.85	11/16	100.5	14.15	11.06
3/4	44.58	2.8	2.2	3/4	104.1	15.18	11.88
13/16	48.28	3.28	2.58	13/16	107.8	16.30	12.76
7/8	52.12	3.8	2.98	7/8	112.3	17.45	13.66
15/16	56 05	4.35	3.41	15/16	116.	18.10	14.61
1	59.48	4.95	3.9	2	119.6	19.78	15.58

Head Linings.

Vose, Dinsmore & Co........New York........See Page 60

Heating Apparatus

A. Carr........New York........See Page 244

Hinge Nails.

D. Brewer & Co........Philadelphia, Pa........See Page 246

Hinges.

Lewis Oliver & Phillips; H. B. Newhall, Ag't...New York........See Page 166
Pratt & Co........Buffalo, N. Y........ " 132
Providence Tool Co; H. B. Newhall, Agent....New York........ " 166
Wm. Green & Co........Wilmington, Del........ " 248

Hoes.

Hermann Boker & Co........New York........See Pages 152, 154

Hoisting Grapple.

Hermann Boker & Co........New York........See Pages 152, 154

Hoisting Machines.

J. S. Mundy........Newark, N. J........See Page 266
R. L. Howard & Son........Buffalo, N. Y........ " 40
Skinner & Gifford Manufacturing Co........Dunkirk. N. Y........ " 140
Speedwell Iron Works........New York........ " 122
Wm. Sellers & Co........Philadelphia, Pa........ " 64

Hoop Iron.

Atwater, Wheeler & Co........New Haven, Conn........See Page 10
Jones & Laughlins........Pittsburgh, Pa........ " 32
Marshall Lefferts, Jr—(Galvanized Iron)........New York........ " 44

Hoppers.

L G. Tillotson & Co........New York........See Page 94

Horse Shoe Iron.

Pittsburgh Bolt Co........Pittsburgh, Pa........See Page 118

IRON, SPLICES AND BOLTS REQUIRED FOR ONE MILE OF TRACK.

Tons of Iron.

Rule.—To find the No. of Tons of Rail to the mile, divide the weight per yard by 7, and multiply by 11, thus: for 56 lb. rail, divide 56 by 7, equal 8, multiplied by 11, equal 88 tons, for one mile of single track,

Weight of Rail, per Yard	Tons per Mile.	Weight of Rail, per Yard.	Tons per Mile.
12 pounds.	12 Tons 920 Pounds.	45 pounds.	70 Tons 1600 Pounds.
14 "	22 "	48 "	75 " 960 "
16 "	25 " 320 "	50 "	78 " 1280 "
18 "	28 " 640 "	52 "	81 " 1600 "
20 "	31 " 960 "	56 "	88 "
22 "	34 " 1280 "	57 "	89 " 1280 "
25 "	39 " 6 0 "	60 "	94 " 640 "
26 "	40 " 1920 "	62 "	97 " 960 "
27 "	42 " 960 "	64 "	100 " 1280 "
28 "	44 "	65 "	102 " 320 "
30 "	47 " 320 "	68 "	106 " 1920 "
33 "	51 " 1920 "	70 "	110 "
35 "	55 "	72 "	113 " 320 "
40 "	62 " 1920 "	76 "	119 " 960 "

Splices and Bolts for One Mile of Track.

30	Feet of Rail,	requires	704	Splices;	1,408	Bolts and Nuts.
28	"	"	754	"	1 508	" "
27	"	"	782	"	1,564	" "
25	"	"	844	"	1,688	" "
24	"	"	880	"	1,760	" "

NUMBER OF RAILS, CHAIRS, OR JOINTS, PER MILE OF SINGLE TRACK.

Length of Rail.	No.	Length of Rail.	No.
18 Feet.	587	24 Feet.	440
20 "	528	26 "	406
21 "	503	28 "	377
22 "	480	30 "	352

NUMBER OF CROSS-TIES PER MILE OF SINGLE TRACK.

Distance from centre to centre,		
Distance from centre to centre,	2 feet	2641 ties.
" " "	2¼ "	2348 "
" " "	2½ "	2113 "
" " "	2¾ "	1921 "
" " "	3 "	1761 "

SPIKES TO THE MILES, SINGLE TRACK.

Sizes,		
Sizes,	5½ by ½	5000 lbs. equals 33 1-3 kegs of 150 lbs. each.
"	5½ by 9/16	4230 lbs. " 28 1-5 " 150 "
"	5½ by ½	about 342 to a keg of 150 lbs.
"	5½ by 9/16	" 284 " " 150 lbs.

IMPROVED TRACING PAPER.

Puscher of Nuremburg has lately suggested a solution of castor oil in absolute alcohol for the purpose of manufacturing a tracing paper. The oil is to be diluted with one, two, or three times its bulk of alcohol, according to the thickness of the paper and the amount consequently required for rendering it transparent. This can be laid on by means of a sponge; and in a very few minutes after the application the paper will be dry, transparent, and ready for use. It will readily receive the mark of a pencil or India ink, and as by immersion in absolute alcohol the oil can be removed, the paper can be restored to its original condition, if desired.

Hydrants.

A. Carr....New York....See Page 244
R. A. Brick & Co.... " " .... " 48

Hydraulic Engines.

Winans & Co....New York....See Page 106

Hydraulic Jacks.

L. G. Tillotson & Co....New York....See Page 94
R. Dudgeon.... " " .... " 70
Vose, Dinsmore & Co.... " " .... " 60

Hydraulic Machinery.

Bolen, Crane & Co....Newark, N. J....See Page 62
Winans & Co....New York.... " 106

Hydraulic Pipe.

Winans & Co....New York....See Page 106

Hydraulic Presses.

Winans & Co....New York....See Page 106

Hydraulic Punches.

R. Dudgeon....New York....See Page 70
Winans & Co.... " " .... " 106

Hydraulic Valves.

Winans & Co....New York....See Page 106

Hydrostatic Wheel Presses.

Wm. Sellers & Co....Philadelphia, Pa....See Page 64

Incrustation Powders.

H. N. Winans....New York....See Page 108
Wiard Locomotive Attachment Co (Anti-Incrustation).... " " ....See Page 110

SHEETS OF SHEET-COPPER.

Standard Sizes.

(Ansonia Brass & Copper Co.)

Thickness. B.W.G.	Sizes of Sheets, and Weight of each Sheet in Pounds. 14 by 48	24 by 48	30 by 60	36 by 72	48 by 72
1	—	116	181	261	348
2	—	111	174	250	334
3	—	102	159	230	306
4	—	93	145	209	278
5	—	81	126	182	242
6	—	75	118	169	226
7	—	70	109	157	209
8	—	63	99	142	190
9	—	58	90	130	173
10	—	48	81	117	156
11	—	46	73	104	139
12	—	41	64	91	122
13	—	35	54	78	104
14	—	29	45	65	86
15	—	26	41	59	78
16	—	23	36	52	70
17	—	20	32	45	60
18	—	18	27	39	52
19	—	16	24	35	47
20	—	14	22	32	43
21	—	13	20	29	39
22	6½	12	18	26	35
23	5⅞	10	16	23	31
24	5¼	9	15	21	28
25	4⅝	8	12½	19	25
26	4	7	11	15	21
27	3½	6	9⅜	13	18
28	3	5	7	11	15

Iron.

Chas. W. Matthews..................................Philadelphia, Pa............See Page 38
Jones & Laughlins......................................Pittsburgh, Pa................ " 32
Leng & Ogden..New York... See Page next to Back Cover
Lewis Oliver & Phillips; H. B. Newhall, Ag't...New York....................See Page 166
O. W. Child.. " " .... " 262
P. H. Moffat...Chicago, Ill.................... " 190
Reese, Graff & Woods..................................Pittsburgh, Pa.............. " 68
Tyng & Co...New York...................... " 260

Iron—Bar.

Atwater, Wheeler & Co.................................New Haven, Conn..........See Page 10
Jones & Laughlins..Pittsburgh, Pa.............. " 32
Marshall Lefferts, Jr. (Galvanized Iron)........New York...................... " 44
P. H. Moffat...Chicago, Ill..................... " 190
Pittsburg Bolt Co..Pittsburgh, Pa................ " 118
Tyng & Co..New York....................... " 260
Vose, Dinsmore & Co..................................... " " .... " 60
Wm. Green & Co...Wilmington, Del............ " 248

Iron Buildings.

Noyes & Wines...New York.....................See Page 264

LIST OF CALIBRE AND WEIGHTS OF LEAD PIPES.

Furnished by E. W. Blatchford & Co.

Calibre.		Weight per foot. Lbs.
1/4 inch	tubing	3/8
3/8 inch	aqueduct	1/2
	light	3/4
	medium	1
	strong	1 1/2
	extra strong	2
1/2 inch	aqueduct	5/8
	extra light	3/4
	light	1
	medium	1 1/4
	strong	1 3/4
	extra strong	2...
5/8 inch	aqueduct	3/4
	extra light	1 1/4
	light	1 3/4
	medium	2
	strong	2 1/2
	extra strong	3
3/4 inch	aqueduct	1
	extra light	1 1/2
	light	2
	medium	2 1/4
	strong	3
	extra strong	3 1/2
7/8 inch	aqueduct	1 1/2
	extra light	2
	light	2 1/2
1 inch	aqueduct	1 1/2
	extra light	2
	light	2 1/2
	medium	3 1/4
	strong	4
	extra strong	4 3/4
1 1/4 inch	aqueduct	2
	extra light	2 1/2
	light	3
	medium	3 3/4
	strong	4 3/4
	extra strong	6
1 1/2 inch	aqueduct	3
	extra light	3 1/2
	light	4
	medium	5
	strong	6
	extra strong	7 1/2
1 3/4 inch	extra light	3 3/4
	light	4 1/2
	medium	5 1/2
	strong	6 1/2
	extra strong	8
2 inch	waste	3
	extra light	4
	light	5
	medium	7
	strong	8
	extra strong	9
2 1/2 inch	3/16 thick	8
	1/4 thick	11
	5/16 thick	14
	3/8 thick	17
3 inch	waste	5
	3/16 thick	9
	1/4 thick	12
	5/16 thick	16
	3/8 thick	20
3 1/2 inch	1/4 thick	15
	5/16 thick	18
	3/8 thick	21
4 inch	waste	5
	1/4 thick	16
	5/16 thick	21
	3/8 thick	25
	7/16 thick	30
4 1/2 inch	waste	6
5 inch	waste	8

Aqueduct Pipe is put up on reels. Other sizes in coils of about 170 lbs. each. Waste Pipe, and all pipe over 2 inches in diameter, is made in lengths of 10 to 15 feet.

Iron Burs.

D. Brewer & Co..Philadelphia, Pa............See Page 246

Iron Doors and Shutters.

Noyes & Wines..New York.......................See Page 264

Iron Fittings.

A. Carr ....New York.....................See Page 244
Hart, Ball & Hart..Buffalo, N. Y.................. " 136
J. J. Walworth..Chicago, Ill........................ " 124
McNab & Harlin Manufacturing Co................New York.......................... " 220
Pancoast & Maule..Philadelphia, Pa............ " 26
See also Gas and Steam Fittings.

Iron Founders.

Bowlers, Maher & Brayton..........................Cleveland, Ohio..............See Page 42
Davenport, Fairbairn & Co.......................... " " .... " 188
Jonas S. Heartt & Co......................................Troy, N. Y...................... " 134
Ramapo Wheel & Foundry Co........................Ramapo, N. Y................ " 256
The Pratt & Whitney Co..................................Hartford, Conn.............. " 142
Wason Manufacturing Co................................Springfield, Mass............ " 146
Wm. P. Kellogg & Co......................................Troy, N. Y........................ " 98

Iron Merchants.

Jno. W. Quincy..New York....................See Page 178
Leng & Ogden....................................New York...See Page next to Back Cover
Tyng & Co.. " " ....See Page 260

Iron Ore Crushers.

Phelps & Sanger..Cleveland, Ohio..............See Page 120

GLASS.

Number of Panes per 50 Feet, or in 1 Box.

Size.	Panes.	Size.	Panes.	Size.	Panes.	Size.	Panes.
6 by 8	150	12 by 14	43	14 by 24	22	20 by 24	15
7 " 9	114	12 " 15	41	15 " 15	32	20 " 25	14
8 " 10	90	12 " 16	38	15 " 16	30	20 " 26	14
8 " 11	82	12 " 17	35	15 " 18	27	20 " 28	13
8 " 12	75	12 " 18	33	15 " 20	24	21 " 27	13
9 " 10	80	12 " 19	32	15 " 21	23	22 " 24	14
9 " 11	73	12 " 20	32	15 " 22	22	22 " 26	13
9 " 12	67	12 " 21	29	15 " 24	20	22 " 28	12
9 " 13	61	12 " 22	28	16 " 16	28	24 " 28	11
9 " 14	57	12 " 23	26	16 " 17	26	24 " 30	10
9 " 15	53	12 " 24	25	16 " 18	25	24 " 32	9
9 " 16	50	13 " 14	39	16 " 20	23	25 " 30	10
10 " 10	72	13 " 15	37	16 " 21	21	26 " 36	8
10 " 12	60	13 " 16	35	16 " 22	20	28 " 34	8
10 " 13	55	13 " 17	33	16 " 24	19	30 " 40	6
10 " 14	51	13 " 18	31	17 " 17	25	31 " 36	6
10 " 15	48	13 " 19	29	17 " 18	24	31 " 40	6
10 " 16	45	13 " 20	27	17 " 20	21	31 " 42	6
10 " 17	42	13 " 21	26	17 " 22	19	32 " 42	5
10 " 18	40	13 " 22	25	17 " 24	18	32 " 44	5
11 " 11	59	13 " 24	23	18 " 18	22	33 " 45	5
11 " 12	55	14 " 14	37	18 " 20	20	34 " 46	5
11 " 13	50	14 " 15	34	18 " 22	18	30 " 52	4
11 " 14	47	14 " 16	32	18 " 24	17	32 " 56	4
11 " 15	44	14 " 17	30	19 " 19	20	36 " 58	3
11 " 16	41	14 " 18	29	19 " 20	19	38 " 58	3
11 " 17	39	14 " 19	27	19 " 22	17	40 " 60	3
11 " 18	36	14 " 20	26	19 " 24	16		
12 " 12	50	14 " 21	24	20 " 20	18		
12 " 13	46	14 " 22	23	20 " 22	16		

LIGHT LOCOMOTIVES

AN EXCLUSIVE SPECIALTY.

Outside Connected Mine Locomotive, 66 inches High, and 65 inches Wide on 3 feet Gauge.

NARROW GAUGE PASSENGER, with 4 drivers and two wheel truck.		With 6 wheel tender, 10 x 16 Cylinders.	With 4 wheel tender, 8 x 16 Cylinders.
NARROW GAUGE FREIGHT 6 drivers, with or without truck; saddle or tender-tank.		12 x 16, 11 x 16,	10 x 16, $9\frac{1}{4}$ x 14.
SPECIAL ENGINES for R. R. Construction and Shifting, Mills, Furnaces, Quarries, Lumber Roads, with Wooden or Iron Rails, Contractors' Use, &c.	With saddle or tender tank, 4 or 6 drivers.	12 x 18, 10 x 16, $9\frac{1}{2}$ x 14, $8\frac{1}{4}$ x 16, 8 x 14,	12 x 16, 9 x 16, $9\frac{1}{4}$ x 14, 8 x 16, 7 x 12,
MINING ENGINES,		Outside connected as above, or inside connected, 9 x 12.	

Photograph and Price of Engine, to do the required work, furnished on application. Inquiry should state length of road, gauge, steepest and average grades, radii of curves and whether on grades, weight and kind of rail, and load to be hauled.

PORTER, BELL & CO.,

PITTSBURGH, PENN'A.

Works, 50th St. & A. V. R. R   Office, 5 Monongahela House.

DIAGRAM showing proportion between

Maximum Load of Locomotive on Level and on Grades

up to 150 feet per mile. Furnished by *Porter, Bell & Co., Pittsburgh, Pa*

See opposite page.

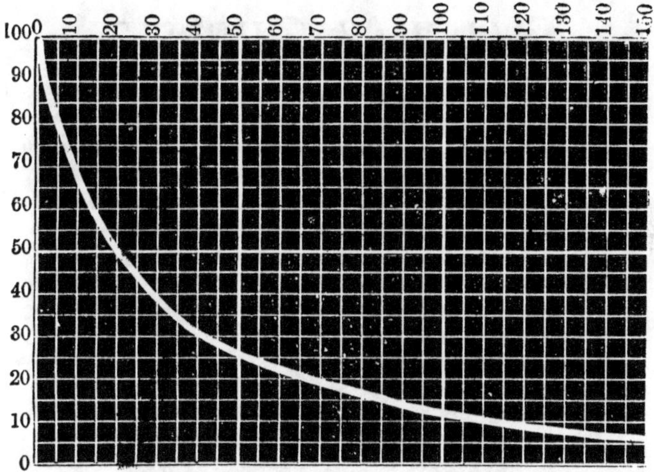

Example I.—What proportion of its load on level will engine haul on grade of 20 feet per mile?

Look for 20 foot grade at top, follow down perpendicular line to curve, and for percentage look at left hand on horizontal line intersected by the curve.

Answer.—50 per cent.

Example II.—If an engine's maximum load on 20 foot grade is 200 tons, what is its capacity on 40 foot grade?

The diagram gives engine on 20 foot grade 50 per cent. of its capacity on level. Then, 200 tons being 50 per cent., the maximum load on level is 400 tons. The percentage shown on 40 foot grade is 30, and 30 per cent. of 400 gives

Answer, 120 tons.

This Diagram supposes track to be straight and in good order; the load is to be started on grade; the level is supposed to be absolute, though 5 or 10 feet to the mile is as near a level as is often found in practice. Grades of 500 feet per mile can be overcome by engines working by adhesion alone on good **T** rail. Tenders must be reckoned as a part of the train.

Iron Ores.

Iron Padlocks.

Iron Paints.

Iron—Pig.

See Pig Iron.

WHITEWASH.

For Outside Wood Work.

Slack ½ bushel of lime, add 2 lbs of white vitrol and water enough to bring to consistence of thick whitewash. This is white and may be colored by adding ochre, umber, lamp black, etc. When lamp black is added to water colors it should first be dissolved in alcohol.

For Stone or Brick Work, etc.

Slack ½ bushel of lime, then fill the barrel two-thirds full of water and add 1 bushel of hydraulic cement, also 3 lbs. of white vitrol dissolved in water. Make about the thickness of paint and lay on with a whitewash brush. It may be improved by adding a peck of white sand before using. This mixture may be colored same as the preceeding one.

For Inside Work.

Whitewash for inside work shonld have no salt in it, for making it permanent use about a quart of thin glue water, or any good clear *size*, to a pailful of wash.

NUMBER OF NAILS AND TACKS PER POUND.

NAILS.

Title.		Size.		No. per lb.
3-penny	fine	1⅛	inch.	760 nails.
3 "	......	1¼	"	480 "
4 "	......	1½	"	300 "
5 "	......	1¾	"	200 "
6 "	......	2	"	160 "
7 "	......	2¼	"	128 "
8 "	......	2½	"	92 "
9 "	......	2¾	"	72 "
10 "	......	3	"	60 "
12 "	......	3¼	"	44 "
16 "	......	3½	"	32 "
20 "	......	4	"	24 "
30 "	......	4½	"	18 "
40 "	......	5	"	14 "
50 "	......	5½	"	12 "
60 "	......	6	"	10 "
6 "	fence,	2	"	80 "
8 "	"	2½	"	50 "
10 "	"	3	"	34 "
12 "	"	3¼	"	29 "

TACKS.

Title.	Length.		No. per lb.
1 oz.	⅛	inch.	16,000
1½ "	3/16	"	10,666
2 "	¼	"	8,000
2½ "	5/16	"	6,400
3 "	⅜	"	5,333
4 "	7/16	"	4,000
6 "	9/16	"	2,666
8 "	⅝	"	2,000
10 "	11/16	"	1,600
12 "	¾	"	1,333
14 "	13/16	"	1,143
16 "	⅞	"	1,000
18 "	15/16	"	888
20 "	1	"	800
22 "	1 1/16	"	727
24 "	1⅛	"	666

PASTE, THAT WILL ADHERE TO ANY SUBSTANCE.

Sugar of lead, 720 grs., and alum, 720 grs.; both are dissolved in water. Take 2½ ounces of gum arabic and dissolve in 2 quarts of warm water. Mix in a dish 1 pound of wheat flour with the gum water cold till in pasty consistence. Put the dish on the fire, pour into it the mixture of alum and sugar of lead. Shake well, and take it off the fire when it shows signs of ebulition. Let the whole cool, and the paste is made. If the paste is too thick, add to it some gum water, till in proper consistence.

Iron Pipe.

A. Carr....New York....See Page 244
Evans, Dalzell & Co....Pittsburgh, Pa.... " 58
Hart, Ball & Hart....Buffalo, N. Y.... " 136
J. D. West & Co....New York.... " 56
McNab & Harlin Manufacturing Co.... " " .... " 220
Morris, Tasker & Co....Philadelphia, Pa.... " 36
Phelps & Sanger....Cleveland. Ohio.... " 120

Iron—Railroad.

See Railroad Iron.

Iron Rods.

Greene & Randolph....New York....See Page 54
Marshall Lefferts, Jr. (Galvanized Iron).... " " .... " 41
Rhode Island Nut Co; H. B. Newhall, Agent... " " .... " 166

Iron—Sheet.

See Sheet Iron.

Iron Shutters and Doors.

Noyes & Wines....New York....See Page 264

PRUSSIAN RAILWAYS.—2 PAGES.

Regulations for Construction and Rolling Stock.

RAILWAY 4 FEET 8½ INCH GAUGE.

Minimum breadth of roadway, double line.... 24 ft. 9 in.
" " single line.... 15 ft. 6 in.
Maximum gradient, level districts.... 1 in 200.
" hilly districts.... 1 in 100.
" mountainous districts.... 1 in 40.
Minimum radius of curves, flat districts.... 3,600 feet.
" " hilly districts.... 1,200 "
" " mountainous districts.... 600 "
Maximum guage for straight lines and curves of more than 1,000 feet radius.... 4 ft. 8½ in.
Maximum guage for round sharp curves.... 4 ft. 9½ in.
Minimum width of top table of rails.... 2¼ inches.
Radius of " " from.... 7 to 5 in.
Minimum depth of rails.... 4½ inches.
Inclination of rails inwards.... 1 in 20.
Minimum depth of ballast under sleepers.... 8 inches.

LEVEL CROSSINGS AND STATIONS.

Minimum angle of crossing, with rails.... 30°.
" width of channel for flange at crossings.... 2⅝ inches.
" depth of " " " .... 1½ "
" portion of level at stations, flat districts.... 1,800 feet.
" " " mountainous do.... 600 "
" distance, centre to centre of rail in station.... 14 "
" radius of crossings and turnouts.... 600 "
" " for facing points at stations for through trains.... 1,000 "
" diameter of engine turn-tables.... 38 "
" width of passenger platforms.... 18 "
Maximum height.... 18 inches.
Minimum distance of pillars in stations from rails.... 9 ft. 5 in.
" diameter of water-crane pipes.... 6 inches.
" height of fixed portion of water-crane above rails.... 9 ft. 6 in.

LOCOMOTIVE SHEDS.

Engine pits, in depth, from.... 2 ft. 6 in. to 3 ft. 6 in.
Minimum height of water tanks above rails.... 17 feet.
" " engine doorways.... 15 ft. 9 in.
" width of " .... 11 feet.
" height of the beams from rails.... 19 "

CARRIAGE SHEDS.

Minimum distance of rail, centre to centre.... 14 ft. 6 in.
" height of doorways.... 13 feet.
" width of " .... 11 "

JONAS S. HEARTT & CO.

MANUFACTURERS OF

RAILROAD CAR AND ENGINE WHEELS.

STREET CAR WHEELS,

C. S. BOSWORTH'S PATENT—Best in Use.

☞AXLES FURNISHED AND FITTED.

TROY, NEW YORK.

AGENTS FOR SALE OF

GEO. W. SWETT & CO'S.

Shaftsbury Charcoal Pig Iron.

PHILADELPHIA SMELTING COMPANY

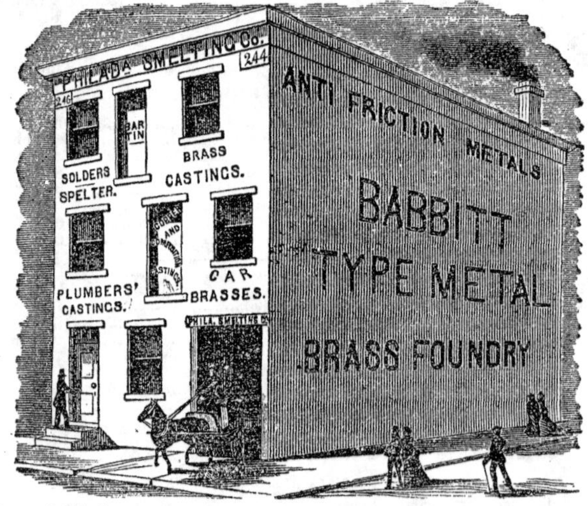

ffice and Works, 244 & 246 South 21st Street, PHILADELPHIA.

Iron Tubing.

Iron Wire.

PRUSSIAN RAILROADS—CONTINUED.

LOCOMOTIVES, SPEED, &c.

Maximum wheel base of engines, with curves 2,000 ft. radius	15 feet.
" " " " 1,500 "	13 "
" " " " 1,000 "	11 "
Maximum load on one axle	12½ tons.
Minimum play of flanges of wheels	⅜ inches.
" clear space between wheels	4 ft. 5½ in.
Maximum projection of flange below top of rail	1¼ inches.
" width of tyre	6 inches.
Minimum "	5¼ inches.
" diameter of driving wheel for goods engine	4 feet.
" " " " passengers' engine	5 "
" " " " express engine	6 "
" " leading and trailing wheel	3 "
Speed of goods train, per hour	16 miles.
Passenger train, per hour	from 25 to 33 miles.
Express train, per hour	40 miles.
Maximum pressure in boiler, per square inch	100 lbs.
Proof pressure in testing boilers	150 lbs.
Height of life guards above rails	2 to 2½ in.
Maximum breadth of engine	10 feet.
" height of chimney above rails	15 "
Minimum diameter of tender wheels	3 "
Maximum breadth of tender	9 "
" height of tender above rails	8 "

CARRIAGES AND WAGONS.

Maximum wheel base for curves of 2,000 feet radius	18 feet.
Inclination of cone of tyres	1-20.
Maximum breadth of tyre	6 inches.
Minimum " "	5 "
" thickness of tyre over rail	¾ "
" play for flanges	⅜ "
Maximum "	1 inch.
Clear width between wheels	4 ft. 5½ in.
Maximum projection of flange from surface of rail	1¼ inches.
Minimum diameter of wheels	3 feet.
" " axle at nave, with 5 tons load per axle	4½ inches.
" " axle journal, with 5 tons load	3 "
Length of axle, centre to centre of journal	6 ft. 5½ in.
Maximum length of journal	8 inches.
Minimum " "	5 "
Height of centre of buffers above rails	3 ft. 5 in.
Distance of centre of buffers apart	5 ft. 9 in.
" side chains apart	3 ft. 6 in.
Minimum diameter of buffers	14 inches.
" rounded projection	1 inch.
Maximum breadth of carriage	8 ft. 7 in.
" " wagons	7 feet.
Maximum height of carriages and wagons	12 ft. 4 in.

Iron Working Machinery.

Bolen, Crane & Co	Newark, N. J	See Page 62
Buffalo Machinery Agency	Buffalo, N. Y	" 160
Isaac H. Shearman	Philadelphia, Pa	" 80
L. W. Pond	Worcester, Mass	" 158
Phelps & Sanger	Cleveland, Ohio	" 120
The Atlas Works	Pittsburgh, Pa	" 172
The Pratt & Whitney Co	Hartford, Conn	" 142
Thorne, DeHaven & Co	Philadelphia, Pa	" 38
Wm. B. Bement & Son	" "	" 168
Wm. Sellers & Co	" "	" 64
Winans & Co	New York	" 106

Japans.

Edward Smith & Co	New York	See Page 218
Valentine & Co	" "	" 198

Joints.

Pittsburgh Bolt Co	Pittsburgh, Pa	See Page 118

MEMORANDA ON ENGLISH RAILWAYS.—2 PAGES.

Parliamentary Regulations for Railways Crossing Roads.

	Turnpike Road.		Public Road.		Occupation Road.	
	ft.	in.	ft.	in.	ft.	in.
Clear width of under bridge or approach	35	0	25	0	42	0
Clear heigth of underbridge for a width of 12 feet	16	0	—	–	—	–
Do., for a width of 10 feet	—	–	45	0	—	–
Do., for a width of 9 feet	—	–	—	–	44	0
Do., at springing	12	0	42	0	—	–
Over bridge. Height of parapets	4	0	4	0	4	0
Approaches. Inclination	4 in 30		4 in 20		4 in 46	
Approaches. Height of fencing	3	0	3	0	3	0

	Narrow.		Broad.	
	ft.	in.	ft.	in.
Average formation width, single line	48	0	24	6
" " " double line	30	0	38	0
Average width of top ballast, single line	43	6	45	6
" " " double line	24	6	29	0

Formation slope from centre in cuttings, 1 in 30.
Width of land taken beyond bottom o slope, from 9 to 12 feet.
Ditch with slopes 1 to 1, 1 foot wide at bottom.
Quick mound, 1 foot 6 inches high.
Post and rail fencing posts, 9 feet apart, 3 feet in the ground. 4 rails, $4 \times 1\frac{1}{2}$.
Dimensions of do., 7 feet 6 inches $\times$ 6 inches $\times 3\frac{1}{2}$.
Intermediate posts, 3 feet apart, 5 feet 6 inches $\times 4 \times 1\frac{1}{2}$.

Board of Trade Regulations for Railways.

Main Line.

Breaking weighth of cast-iron girders to be made = permanent load $\times 3 +$ moving load $\times 6$.

Wrought-iron bridges must not be strained to more than 5 tons per square inch when loaded with the heaviest engines.

Minimum distance of standing work from the outer edge of rail at level of carriage steps = 3 feet 6 inches in England = 4 feet in Ireland.

Minimum distance between lines of railway = 6 feet.

Stations.

Minimum width of platforms = 6 feet.
" distance of columns from edge of platform = 6 feet.
Steepest gradient recommended for stations = 1 in 300.

Carriages.

Minimum space per passenger = 20 cubic feet.
" area of glass per passenger = 60 superficial inches.
" width of seats = 15 inches.
" breadth of seat per passenger = 16 inches.
" number of lamps per carriage = 2.

Journal Bearings.

G. W. W. Stevens & Co..........New York..........See Page 8
John Harden.......... " " .... " 230

Knife Grinders.

D. Brewer & Co..........Philadelphia, Pa..........See Page 246

Lace Leather.

Bickford, Curtiss & Deming..........Buffalo, N. Y..........See Page 136
N. H. Gardner & Co.......... " " .... " 254

Ladles.

Wm. B. Bement & Son..........Philadelphia, Pa.......... See Page 168

Lag Screws.

Lewis Oliver & Phillips; H. B. Newhall, Agt...New York..........See Page 166
Pittsburgh Bolt Co..........Pittsburgh, N. Y.......... " 118
Providence Tool Co.; H. B. Newhall, Agent....New York.......... " 166

Lamp Posts.

Morris, Tasker & Co..........Philadelphia, Pa..........See Page 36
R. A. Brick & Co..........New York.......... " 48

MEMORANDA ON ENGLISH RAILWAYS—Continued.

Requirements.

Joints of rails to be fished.
Chairs to be secured by iron spikes.
Fang bolts to be used at the joints of flat-bottomed rails.

Stations.

Ends of platforms to be ramped (not *stepped*).
Signals and distant signals in both directions.
Signals to be weighted to fly to "danger" if wire breaks.
Switch handles to be brought together.
" not to be worked between lines of rail.
Facing points to be avoided
" where unavoidable, to be furnished with self-acting signal.
Sidings to be supplied with locked chock-block.
Sidings on a gradient falling towards the line to be provided with a blind siding.
Turn-tables to be a safe distance from adjacent lines.
Platforms of bridges protected from fire.
Walls and handrails to be provided on viaducts at stations.
Level crossing gates to be capable of being closed across both road and railway.
Clocks visible at all stations.
Mile posts and gradient boards.
Tunnels and dangerous places protected by telegraph.

RECIPES FOR SEALING WAX.

Red.

Shellac	4	ounces.
Venice turpentine	1¼	"
Vermilion	3	"

Black.

Shellac	60	parts.
Fine ivory black	30	"
Venice turpentine	30	"

Yellow.

Shellac	4	ounces.
Resin	1¼	"
Venice turpentine	2	"
Sulphuret of arsenic	¼	"

Brown.

Shellac	7½	ounces.
Venice turpentine	4	"
Brown ochre	1	"
Vermilion	½	"

Blue.

Fine shellac	7	ounces.
Venice turpentine	3	"
Resin	1	"
Mineral Blue	1	"

Green.

Shellac	4	ounces.
Venice turpentine	2	"
Resin	1¼	"
Sulphuret of arsenic	¼	"
Mineral blue	¼	"

Melt the shellac at the lowest necessary temperature; stir the turpentine previously warmed; and then add the colors, and pour into moulds

Lamps.

Ansonia Brass & Copper Co	New York	See Page 122
Buffalo Steam Gauge Co	Buffalo, N. Y	" 170
J. G. Knapp Manufacturing Co	New York	" 16
L. G. Tillotson & Co	" "	" 94
S. M. Aikman & Co	" "	" 114
Vose, Dinsmore & Co	" "	" 60

Lanterns.

Buffalo Steam Gauge C	Buffalo, N. Y	See Page 170
L. G. Tillotson & Co	New York	" 94
Parmelee & Bonnell	Buffalo, N. Y	" 202
Orrin L. Gridley	" "	" 196
Radley & McAlister Manufacturing Co	New York	" 250
S. M. Aikman & Co	" "	" 114
T. B. Bickerton & Co	Philadelphia, Pa	" 138
Vose, Dinsmore & Co	New York	" 60

STANDARD SIZES OF NUTS.

Square Nuts.

Width.	Thickness.	Hole.	Size of Bolt.
1/2	1/4	7-32	1/4
5/8	5/16	9-32	5/16
3/4	3/8	11-32	3/8
7/8	7/16	13-32	7/16
7/8	1/2	7/16	1/2
1	1/2	7/16	1/2
1 1/8	1/2	1/2	9/16
1 1/8	5/8	9/16	5/8
1 1/4	5/8	9/16	5/8
1 3/8	3/4	11/16	3/4
1 1/2	3/4	11/16	3/4
1 5/8	7/8	13/16	7/8
1 3/4	7/8	13/16	7/8
1 3/4	1	29-32	1
2	1	29-32	1
2	1 1/8	1	1 1/8
2 1/4	1 1/8	1	1 1/8
2 1/4	1 3/8	1 1/8	1 1/4
2 1/2	1 1/4	1 1/8	1 1/4
2 3/4	1 3/8	1 1/4	1 3/8
3	1 1/2	1 3/8	1 1/2
3 1/4	1 5/8	1 7/16	1 5/8
3 1/2	1 3/4	1 9/16	1 3/4
3 3/4	1 7/8	1 11/16	1 7/8
4	2	1 13/16	2

Hexagon Nuts.

Width.	Thickness.	Hole.	Size of Bolt.
7/8	3/8	11-29	3/8
3/4	1/2	7/16	1/2
1	1/2	7/16	1/2
1 1/8	5/8	9/16	5/8
1 1/4	3/4	9/16	5/8
1 3/8	3/4	11/16	3/4
1 3/8	7/8	11/16	3/4
1 5/8	7/8	13/16	7/8
1 5/8	1	13/16	7/8
1 3/4	1	29-32	1
1 3/4	1 1/8	29-32	1
2	1 1/4	1	1 1/8
2 1/4	1 3/8	1 1/8	1 1/4
2 1/2	1 1/2	1 1/4	1 3/8
2 3/4	1 5/8	1 3/8	1 1/2
3	1 3/4	1 7/16	1 5/8
3 1/4	1 7/8	1 9/16	1 3/4
3 1/2	2	1 11/16	1 7/8
3 1/2	2	1 13/16	2

NUMBER OF NUTS OF EACH SIZE TO 100 POUNDS.

Cold Punched Nuts.

	Square.	Hexagon.
3/8-in.	1945	3010
1/2 "	810	810
5/8 "	425	450
3/4 "	243	265
7/8 "	166	160
1 "	110	105
1 1/8 "	80	80
1 1/4 "	67	60
1 1/2 "	32	30

Hot Pressed Nuts.

	Square.	Hexagon.
1/2-in.	1090	1240
5/8 "	540	645
3/4 "	370	420
7/8 "	225	253
1 "	145	165
1 1/8 "	95	115
1 1/4 "	68	83
1 1/2 "	43	50

STANDARD LIGHT SIZES, HOT PRESSED, SQUARE AND HEXAGON NUTS.

Short Diam.	Thickness.	Hole.	Size of Bolt.
5/8	3/8	21-64	3/8
3/4	7/16	3/8	7/16
7/8	1/2	7/16	1/2
1	9/16	1/2	9/16
1 1/8	5/8	9/16	5/8
1 1/4	3/4	21-32	3/4

Short Diam.	Thickness.	Hole.	Size of Bolt.
1 1/2	7/8	49-64	7/8
1 3/4	1	7/8	1
2	1 1/8	1	1 1/8
2 1/4	1 1/4	1 3-32	1 1/4
2 1/2	1 3/8	1 7-32	1 3/8
2 3/4	1 1/2	1 5/16	1 1/2

THE PRATT & WHITNEY CO.,

HARTFORD, CONN.,

MANUFACTURERS OF

Engine Lathes, With Slate's Patent Taper Attachment.

Boring Mills, Shaping Machines,

Planers, Drills,

AND OTHER *Machinists' Tools,*

Gun and Sewing Machine Machinery, Friction Clutch Pulleys, &c.

☞ SEND FOR DESCRIPTIVE PAMPHLET. ☜

NEW HAVEN NUT CO.

MANUFACTURERS OF

PATENT

HOT PRESSED NUTS,

P. O. ADDRESS,

WESTVILLE, CONN.

L. P. WOODWORTH,
Superintendent.

J. D. PAYNE,
S c'y and Treas.

THREE FIRE ENGINES

AT A BARGAIN.

One Steamer, fitted for Horses, Prince's Metal Jacket, very powerful and highly finished.

One Steamer, fitted for Horses, Sheet Iron Jacket.

One Single Pump Steamer, drawn by Hand, Sheet Iron Jacket.

☞ All built by *The Allerton Iron Works Manufacturing Co.*, and will be sold at very low prices.

ADDRESS GEORGE M. ALLERTON,

205 BROADWAY, NEW YORK.

Lard Oil.

F. S. Pease................Buffalo, N. Y................See Page 100
P. H. Moffat................Chicago, Ill................ " 190

Lathe Centre Grinders.

D. Brewer & Co................Philadelphia, Pa................See Page 246

Lathe Chucks.

D. Brewer & Co................Philadelphia, Pa................See Page 246

Lathe Dogs.

Champlin & Rogers................Chicago, Ill................See Page 192
The Worcester Machine Screw Co.; H. B. Newhall, Agent................New York................See Page 166

Lathes.

Bolen, Crane & Co................Newark, N. J................See Page 62
Isaac H. Shearman................Philadelphia, Pa................ " 80
L. W. Pond................Worcester, Mass................ " 158
Phelps & Sanger................Cleveland, Ohio................ " 120
The Buffalo Machinery Agency................Buffalo, N. Y................ " 160
The Pratt & Whitney Co................Hartford, Conn................ " 142
Wm. B. Bement & Son................Philadelphia, Pa................ " 168
Wm. Sellers & Co................ " " .... " 64

Lead.

Bruce & Cook................New York................See Page 112
E. W. Blatchford & Co................Chicago, Ill................ " 124
Holmes & Lissberger................New York................ " 84
Philadelphia Smelting Co................Philadelphia, Pa................ " 134

STANDARD SIZES OF WASHERS.

Diam.	Size of Hole.	Thickness, Wire Guage.	Size of Bolt.	Diam.	Size of Hole.	Thickness, Wire Guage.	Size of Bolt.
1/2	1/4	No. 18	3/16	1 3/4	11/16	No. 10	5/8
5/8	5/16	" 18	1/4	2	13/16	" 10	3/4
3/4	5/16	" 16	1/4	2 1/4	15/16	" 9	7/8
7/8	3/8	" 16	5/16	2 1/2	1 1/16	" 9	1
1	7/16	" 14	3/8	2 3/4	1 1/4	" 9	1 1/8
1 1/4	1/2	" 14	7/16	3	1 3/8	" 9	1 1/4
1 3/8	9/16	" 12	1/2	3 1/2	1 1/2	" 9	1 3/8
1 1/2	5/8	" 12	9/16				

NUMBER OF WASHERS IN 100 POUNDS.

Size.	No. to 100 lbs.	Size.	No. to 100 lbs.
3/8	5715	1	910
1/2	4010	1 1/8	510
5/8	2235	1 1/4	415
3/4	1450	1 1/2	335
7/8	1185		

TEMPERING STEEL. (*Haswell.*)

Steel in its hardest state being too brittle for most purposes, the requisite strength and elasticity are obtained by tempering—or *letting down the temper*, as it is termed—which is performed by heating the hardened steel to a certain degree and cooling it quickly. The requisite heat is usually ascertained by the color which the surface of the steel assumes from the film of oxide thus formed. The degrees of heat to which these several colors correspond are as follows:

At 430°, a very faint yellow At 450°, a pale straw color..	Suitable for hard instruments; as hammer-faces, drills, &c.
At 470°, a full yellow.......... At 490°, a brown color........	For instruments requiring hard edges without elasticity; as shears, scissors, turning tools, &c.
At 510°, brown, with purple spots.......... At 538°, purple..........	For tools for cutting wood and soft metals; such as plane-irons, knives, &c.
At 550°, dark blue.......... At 560°, full blue..........	For tools requiring strong edges without extreme hardness; as cold-chisels, axes, cutlery, &c.
At 600°, grayish-blue, verging on black..........	For spring-temper, which will bend before breaking; as saws, sword-blades, &c.

If the steel is heated higher than this, the effect of the hardening process is destroyed.

Lead Pipe.

Bruce & Cook..New York....................See Page 112
E. W. Blatchford & Co..................................Chicago, Ill.................... " 124

Learners' Telegraph Instruments.

M. A. Buell..Cleveland, Ohio..............See Page 46

ANGLE AND T IRON.—2 PAGES.

Weight of Running Foot in Pounds.

Weights of ⊔ (channel) and H iron may be found from the Table of Angle and T iron in the following manner, provided the web and flanges are of the same mean thickness:

Let the number of the side column referred to equal half the sum of the depth of web and breadth of both flanges; then twice the weight corresponding thereto will be the weight per lineal foot, according to the thickness, of the channel or H iron required.

Sum of width of Flanges.	Thickness in Fractions of an Inch.									
Ins.	3/16	1/4	5/16	3/8	7/16	1/2	9/16	5/8	11/16	3/4
2	1.13	1.46	1.76	2.03						
1/8	1.21	1.56	1.89	2.19						
1/4	1.29	1.67	2.02	2.34						
3/8	1.37	1.77	2.15	2.5						
1/2	1.45	1.88	2.28	2.66	3.01					
5/8	1.52	1.98	2.41	2.81	3.19					
3/4	1.6	2.08	2.54	2.92	3.37					
7/8	1.68	2.19	2.67	3.13	3.55					
3	1.76	2.29	2.8	3.28	3 74	4.17				
1/8	1.84	2.4	2.93	3.44	3.92	4.37				
1/4	1.91	2.5	3.06	3.59	4.1	4.58				
3/8	1.99	2.6	3.19	3.75	4.28	4.79				
1/2	2.07	2.71	3.32	3 91	4.47	5.	5.51			
5/8	2.15	2.81	3.45	4.06	4.65	5.21	5.74			
3/4	2.23	2.92	3.58	4.22	4.83	5.42	5.98			
7/8	2.3	3 02	3.71	4.38	5.01	5.63	6.21			
4	2.38	3.13	3.84	4.53	5.2	5.83	6.45	7.04		
1/8	2.46	3.23	3.97	4.69	5.38	6.04	6.68	7.29		
1/4	2.54	3.33	4.1	4.84	5.56	6 25	6.91	7.55		
3/8	2.62	3 44	4.23	5.	5.74	6.46	7.15	7.81		
1/2	2.7	3.54	4.36	5.16	5.92	6.67	7.38	8.07		
5/8	2.77	3.65	4.49	5.31	6.11	6.87	7.62	8.33	9 02	
3/4	2.85	3.75	4.62	5.47	6.29	7.08	7.85	8.59	9.31	
7/8	2.93	3.85	4.75	5.62	6.47	7.29	8.09	8.85	9.6	
5	3.01	3.96	4.88	5.78	6.65	7.50	8.32	9.11	9.88	10.62
1/8	3.09	4.06	5.01	5.94	6.84	7.71	8.55	9.37	10.17	10.94
1/4	3.16	4.17	5.14	6.09	7.02	7.91	8.79	9 63	10.45	11.25
3/8	3.24	4.27	5.27	6.25	7.2	8.12	9.02	9.9	10.74	11.56
1/2	3.32	4.37	5.4	6.41	7.38	8.33	9 26	10.16	11.03	11.87
5/8	3.4	4.48	5.53	6.56	7.56	8.54	9.49	10.42	11.31	12.19
3/4	3.48	4.58	5.66	6.72	7.75	8.75	9.73	10.68	11.6	12.5
7/8	3.55	4.69	5.79	6.87	7.93	8.96	9.96	10.94	11.89	12.81
6	3.63	4.79	5.92	7.03	8.11	9.17	10.19	11.2	12.17	13·12
1/8	3.71	4.9	6.05	7.19	8.29	9.37	10.43	11.46	12.46	13.43
1/4	3.79	5.	6.18	7.34	8.48	9.58	10.66	11.72	12.74	13.75
3/8	3.87	5.1	6.32	7.5	8.66	9 79	10.9	11.98	13.03	14.06
1/2	3.95	5.21	6.45	7.66	8.84	10.	11.13	12.24	13.32	14 37
5/8	4.02	5.31	6.58	7.81	9.02	10.21	11.37	12.5	13.6	14.68
3/4	4.1	5.42	6.71	7.97	9.21	10.42	11.6	12.76	13.89	15.
7/8	4.18	5.52	6.84	8.13	9.30	10.63	11.84	13.02	14.18	15.31

Leather Belting.

ANGLE AND T IRON.—CONTINUED.

Weight of Running Foot, in Pounds.

See Note at head of this Table.

Sum of width of Flanges.	Thickness in Fractions of an Inch.									
Ins.	5/16	3/8	7/16	1/2	9/16	5/8	11/16	3/4	7/8	1
7	6.97	8.28	9.57	10.83	12.07	13.28	14.47	15.62	17.86	20.
1/8	7.10	8.44	9.75	11.04	12.3	13.54	14.75	15.93	18.23	20.42
1/4	7.23	8.59	9.93	11.25	12.54	13.8	15.04	16.25	18.59	20.83
3/8	7.36	8.75	10.12	11.46	12.77	14.03	15.33	16.56	18.96	21.25
1/2	7.49	8.91	10.3	11.67	13.01	14.32	15.61	16.87	19.32	21.67
5/8	7.62	9.06	10.48	11.87	13.24	14.58	15.9	17.19	19.69	22.08
3/4	7.75	9.22	10.66	12.08	13.48	14.84	16.18	17.5	21.05	22.5
7/8	7.88	9.37	10.85	12.29	13.71	15.1	16.47	17.81	20.42	22.92
8	8.01	9.53	11.03	12.5	13.94	15.36	16.76	18.12	20.78	23.33
1/8	8.14	9.69	11.21	12.71	14.18	15.62	17.04	18.44	21.15	23.73
1/4	8.27	9.84	11.39	12.92	14.41	15.88	17.33	18.75	21.51	24.16
3/8	8.40	10.	11.57	13.12	14.65	16.15	17.62	19.06	21.87	24.58
1/2	8.53	10.16	11.76	12.33	14.88	16.41	17.9	19.37	22.24	25.
5/8	8.66	10.31	11.94	13.54	15.12	16.67	18.19	19.69	22.6	25.41
3/4	8.79	10.47	12.12	13.75	15.35	16.93	18.47	20.	22.97	25.83
7/8	8.92	10.62	12.3	13.96	15.58	17.19	18.76	20.31	23.33	26.25
9	9.05	10.78	12.49	14.17	15.82	17.45	19.05	20.62	23.7	26.66
1/4	9.31	11.09	12.85	14.58	16.29	17.97	19.62	21.25	24.42	27.50
1/2	9.57	11.41	13.22	15.	16.76	18.49	20.19	21.87	25.15	28.33
3/4	9.83	11.72	13.58	15.42	17.23	19.01	20.77	22.50	25.88	29.16
10	10.09	12.03	13.95	15.83	17.7	19.53	21.34	23.12	26.61	30.
1/4	10.35	12.34	14.31	16.25	18.16	20.05	21.91	23.75	27.34	30.83
1/2	10.61	12.66	14.67	16.67	18.63	20.57	22.49	24.37	28.07	31.67
3/4	10.87	12.97	15.04	17.08	19.1	21.09	23.06	25.	28.8	32.5
11	11.13	13.28	15.4	17.5	19.57	21.61	23.63	25.62	29.53	33.33
1/4	11.39	13.59	15.77	17.92	20.04	22.13	24.2	26.25	30.26	34.17
1/2	11.65	13.91	16.13	18.33	20.51	22.66	24.78	26.87	30.99	35.
3/4	11.91	14.22	16.5	18.75	20.98	23.18	25.35	27.5	31.72	35.83
12	12.17	14.53	16.86	19.17	21.44	23.7	25.92	28.12	32.45	36. 7
1/4		14.84	17.23	19.58	21.91	24.22	26.5	28.75	33.17	37.5
1/2		15.16	17.59	20.	22.38	24.74	27.07	29.37	33.9	38.33
3/4	...		17.96	20.42	22.85	25.26	27.64	30.	34.63	39.16
13	...		18.32	20.83	23.32	25.78	28.22	30.62	35.36	40.
1/4	...	...		21.25	23.79	26.3	28.79	31.25	36.09	40.83
1/2	...	...		21.67	24.26	26.82	29.36	31.87	36.82	41.67
3/4	...	...		22.08	24.73	27.34	29.93	32.5	37.55	42.5
14	...	...		22.5	25.19	27.86	30.51	33.12	38.28	43.33
1/4	...	...		22.92	25.66	28.38	31.08	33.75	39.01	44.17
1/2	...	...		23.33	26.13	28.91	31.65	34.37	39.74	45.
3/4	...	...		23.75	26.6	29.43	32.22	35.	40.47	45.83
15	...	...		24.17	27.07	29.95	32.8	35.62	41.2	46.67

RAILROAD MACHINERY,

TIRES, &c., &c.

JOHN C. ELLIS, *Prest.* CHAS. G. ELLIS, *Treas.* WALTER McQUEEN, *Sup't.*

SCHENECTADY LOCOMOTIVE WORKS,

Schenectady, N. Y.,

Continue to receive orders, and to furnish with promptness the best and latest Improved

Coal or Wood Burning Locomotive Engines.

ALSO

REPAIR AND REBUILD LOCOMOTIVES.

☞ The above Works, located on the N. Y. C. R. R., near the centre of the State, possess superior facilities for forwarding work to any part of the country.

ADHESIVE POWER OF LOCOMOTIVES.

Adhesion per Ton of Load on the Driving Wheels:

When the rails are very dry..................	600 lbs. per ton.
When the rails are very wet..................	550 " "
In misty weather if the rails are greasy..	300 " "
In frosty or snowy weather..................	200 " "

In coupled engines the adhesive force is due to the load on all wheels coupled to the driving wheels.

The adhesive power must exceed the tractive force of an engine on the rails, otherwise the wheels will slip. For loads on driv ng wheels, see below.

DISTRIBUTION OF WEIGHT IN LOCOMOTIVES.

The average distribution of the weights of a six-wheeled locomotive on its wheels is:

Assuming the total weight of the engine in working order to be 1;

	Passenger Engines.		Freight Engines.
Load on leading wheels......	.32	......	.34
" driving wheels......	.48	......	.36
" trailing wheels......	.20	......	.30
Total weight of engine......	100	......	100

Passenger engines, narrow gauge, average...	from 20 to 30 tons.
Freight engines..............................	" 24 to 32 "
Broad-gauge engines, first class..............	" 35 "
Incline engines..............................	" 40 to 47 "

TRACTIVE POWER OF LOCOMOTIVES.

Let D = Diameter of cylinder in inches.
" P = Mean pressure of steam in cylinders in lbs. per square inch.
" L = Length of stroke in inches.
" W = Diameter of driving wheel in inches.

Tractive force on rails in lbs. will equal $\frac{D^2 P L}{W}$.

EFFECTIVE PRESSURE OF STEAM ON PISTON.

With Different Rates of Expansion.—Boiler Pressure being assumed at 100 *lbs. per square inch.*

	Effective pressure.
Steam cut off at ¾ of stroke	= 90
" ⅔ "	= 80
" ½ "	= 69
" ⅓ "	= 50
" ¼ "	= 40

TO FIND THE LOAD WHICH AN ENGINE WILL TAKE ON A GIVEN INCLINE.

Let G = Resistance due to gravity on the steepest gradient in lbs. per ton. See page 151.
" R = Resistance due to assumed velocity of train in lbs. per ton. See page 151.
" T = Tractive power of engine in lbs. as found above.
" W = Weight of engine and tender in tons.

The load the engine can take in tons, including the weight of the waggons, but not that of engine and tender will equal $\frac{T}{G + R} = W$.

PITTSBURGH STEEL WORKS.

ANDERSON & WOODS,

PITTSBURGH, PA.

REVERSIBLE STEEL FROGS,

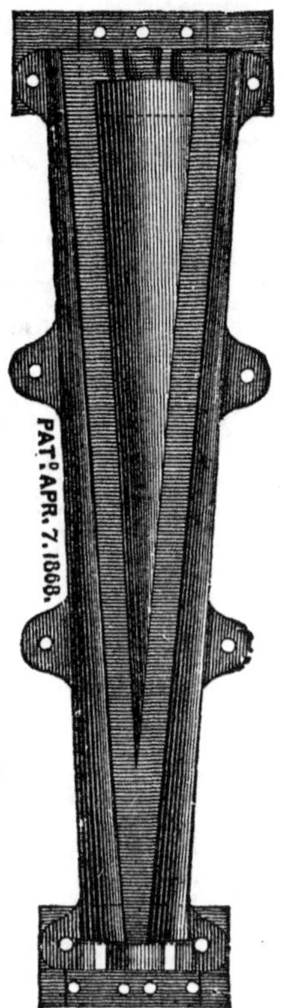

Best Refined Tool Steel,

R. R. Cast Spring Steel,

Frog Points, Side Bars,

Slide Bars, Connecting Rods,

Locomotive and Car Axles,

Crank Pins, Piston Rods,

Cast Steel Crow Bars.

FORGINGS

OF ANY DESCRIPTION MADE TO ORDER.

FROGS AND CROSSINGS

MADE TO ANY ANGLE.

OUR FROGS

ARE NOW IN USE on the FOLLOWING ROADS:

Lake Shore & Michigan Southern Railroad.
New York Central & Hudson River Railroad.
Allegheny Valley Railroad.
Pittsburgh, Washington & Baltimore Railroad.
Wisconsin Central Railroad.
Louisville & Nashville Railroad.
Oil Creek & Allegheny River Railroad.
Union Railway.
Houghton & Ontonagon Railroad.
St. Louis & Iron Mountain Railroad.
Atlantic & Pacific Railroad.

Levels.

F. Eckel.....New York.....See Page 76
Kuebler & Seelhorst.....Philadelphia, Pa..... " 216
Wm. J. Young & Sons..... " "..... " 138

Lightning Rods.

J. D. West & Co.....New York.....See Page 56

Light Locomotives.

Porter, Bell & Co.....Pittsburgh, Pa.....See Page 128

Lime Kiln Blocks.

Philip Neukumet.....Philadelphia, Pa.....See Page 104

Linen Hose.

A. Carr.....New York.....See Page 244

TABLE OF GRADIENTS

AND RESISTANCE DUE TO GRAVITY PER TON, FOR EACH.

(*Haslett.*)

Vertical Rise. Ratio.	Pr. Mile.	Resistance Per Ton.	Vertical Rise. Ratio.	Pr. Mile.	Resistance Per Ton.
one in	feet.	lbs.	one in	feet.	lbs.
100	52.8	22.4	60	88.	37.3
98	53.9	22.8	58	91.	38.6
96	55.	23.3	56	94.2	40.
94	56.1	23.8	54	97.7	41.4
92	57.5	24.3	52	101.5	43.
90	58.6	24.9	50	105.6	44.8
88	60.	25.4	48	110.	46.6
86	61.3	26.	46	115.	48.6
84	62.8	26.6	44	120.	50.9
82	64.3	27.3	42	125.7	53.3
80	66.	28.	40	132.	56.
78	67.6	28.7	38	138.9	58.9
76	69.4	29.4	36	146.6	62.2
74	71.3	32.2	34	155.3	65.8
72	73.3	31.1	32	165.	70.
70	75.4	32.	30	176.	74.6
68	77.6	32.9	28	188.5	80.
66	80.	33.9	26	203.	86.1
64	82.5	35.	24	220.	93.3
62	85.1	36.1	22	240.	101.8

RESISTANCE OF TRAINS, ON A LEVEL.

AT DIFFERENT SPEEDS, IN LBS. per TON OF LOAD.

(*Molesworth.*)

V = Velocity in miles per hour.
R = Resistance in lbs. per ton of train.

$$R = \frac{V^2}{171} + 8.$$

The resistance of curves may be reckoned as 1 per cent. for each degree of the curve occupied by the train.

Imperfections of road vary from 5 to 40 per cent.

Strong side winds..... 20 "

Velocity of train in miles per hour.	10	15	20	30	40	50	60	70
Resistance on straight line in lbs. per ton...	8½	9¼	10¼	13¼	17¼	22½	29	36½
Do., with sharp curves and strong wind*....	13	14	15½	20	26	34	43½	55

* 50 per cent added to resistancc on straight line.

Linseed Cake.

Campbell & Thayer..New York......................See Page 264

Linseed Oil.

Atlantic White Lead Co.—Rob't Colgate & Co. New York...............See Pages 238, 240
Campbell & Thayer.. " " ....See Page 264
C. E. Hecht..Easton, Pa...................... " 158
E. W. Blatchford & Co....................................Chicago, Ill.................... " 124
F. S. Pease...Buffalo, N. Y.................. " 100
John Jewett & Son...New York...................See Pages 72, 74

Litharge.

Atlantic White Lead Co.—Rob't Colgate & Co. New York..............See Pages 238, 240
Brooklyn White Lead Co................................New York.......................See Page 268
Wetherill & Bro...Philadelphia, Pa............. " 52

Lithography.

Allen, Lane & Scott...Philadelphia, Pa............See Page 90
Chas. F. Ketcham..New York...................... " 262
Warren, Johnson & Co ....Buffalo, N. Y.................. " 100

Local and Coupon Tickets.

Allen, Lane & Scott...Philadelphia, Pa............See Page 90
Warren, Johnson & Co.....................................Buffalo, N.Y.................. " 100

SHELLS OF BOILERS.

Resistance to Internal or Bursting Pressure.

(From Haswell.)

Diameter.	Thickness.	Bursting Pressure per Square Inch. Single Riveted.	Double Riveted.	Diameter.	Thickness.	Bursting Pressure per Square Inch. Single Riveted.	Double Riveted.
Feet.	Ins.	Lbs.	Lbs.	Feet.	Ins.	Lbs.	Lbs.
2	1/4	573	745	7.6	5/16	191	248
2.6	1/4	458	596	7.6	3/8	229	298
3	1/4	382	496	8	5/16	179	233
3.4	1/4	318	414	8	3/8	215	279
3.4	5/16	398	518	8.6	5/16	168	219
3.6	1/4	327	426	8.6	3/8	202	263
3.6	5/16	409	532	9	5/16	159	207
4	1/4	286	372	9	3/8	191	248
4	5/16	358	465	9.6	5/16	150	196
4.6	1/4	254	331	9.6	3/8	181	235
4.6	5/16	318	413	10	5/16	143	186
5	1/4	229	298	10	3/8	172	224
5	5/16	286	372	10	1/2	229	298
5.6	1/4	208	270	10.6	5/16	136	177
5.6	5/16	260	338	10.6	3/8	163	212
5.6	3/8	312	406	10.6	1/2	218	284
6	1/4	191	248	11	3/8	156	203
6	5/16	239	311	11	1/2	208	271
6	3/8	286	372	11.6	3/8	149	194
6.6	5/16	220	287	11.6	1/2	199	259
6.6	3/8	264	344	12	3/8	143	166
7	5/16	204	266	12	1/2	191	248
7	3/8	245	319				

Tensile resistance of the plates without Riveting is taken at a mean of 55,000 pounds per square inch.

The single-riveted are estimated at .5 the resistance of the plates, and the staggered riveted at .65.

Such allowances for the resistance and wear of the plates, oxydation, etc., are to be made, as the character of the metal, the nature of the services, and the circumstance of using fresh or salt water, etc., will render necessary.

In riveted plates, it is customary in practice to estimate the safe tensile resistance of the metal of a boiler or tube, when exposed to salt-water, at one-fifth of its ultimate resistance or bursting pressure; and, when exposed to fresh-water alone, at one-fourth of it.

TRENTON VISE and TOOL WORKS,

TRENTON, N. J.

HERMANN BOKER & CO.,

PROPRIETORS.

Office and Warehouse, 101 & 103 Duane Street, and 91 & 93 Thomas Street,

NEW YORK.

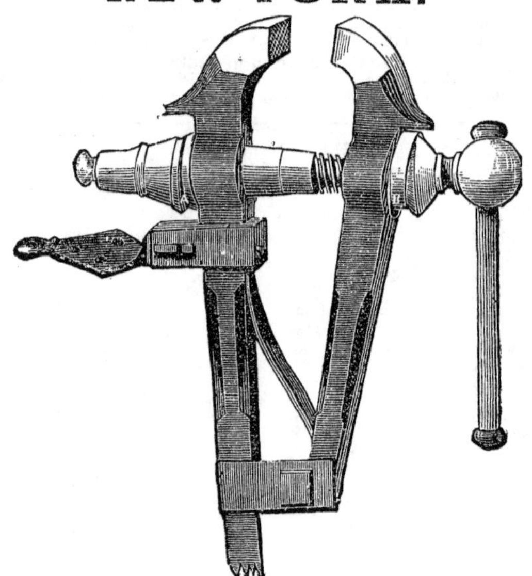

VISES.	HAMMERS.	PICKS.
Swivel Vises.	Striking.	Railroad,
Plain Parallel.	Napping.	Mill.
Coachmakers.	Hand Drilling.	Stone.
Solid Box.	Blacksmiths.	Clay.
Pipe.	Mason.	Coal,
Vise Screws.	Stone.	
	Engineers.	
Stone Sledges.	Chipping.	**Railroad Mauls.**
Smith Sledges.	Riveting.	**Ship Mauls.**
Coal Sledges.	Ball Pene.	**Mattocks.**

Chisels, Tongs, Bench Screws, Wedges, Crowbars, &c.

☞ *SEND FOR ILLUSTRATED PRICE LIST.*

Lock Nuts.

Locks.

Locomotive Balances.

Locomotive Bells.

Locomotive Blocks.

Locomotive Builders.

Locomotive Head Lights.

WROUGHT IRON FLUES.

Resistance to Collapsing Pressure.

(From Haswell.)

Diameter.	Length.	Thickness.	Collapsing Pressure per Square Inch.	Diameter.	Length.	Thickness.	Collapsing Pressure per Square Inch.
Ins.	Feet.	Ins.	Lbs.	Ins.	Feet.	Ins.	Lbs.
6	10	.2	417	11½	12	.2	197
6½	10	.2	385	11½	12	¼	308
7	10	.2	357	11½	12	5/16	532
7	10	¼	580	12	15	.2	153
7½	10	.2	333	12	15	¼	239
7½	10	¼	542	12	15	5/16	415
8	10	.2	312	12½	15	¼	229
8	10	¼	508	12½	15	5/16	398
8½	10	.2	294	13	15	¼	220
8½	10	¼	478	13	15	5/16	384
9	10	.2	278	13½	15	¼	212
9	10	¼	451	13½	15	5/16	369
9½	10	.2	263	14	18	¼	176
9½	10	¼	427	14	18	5/16	305
10	12	.2	227	14½	18	¼	168
10	12	¼	354	14½	18	5/16	294
10	12	5/16	612	15	20	¼	157
10½	12	.2	216	15	20	5/16	276
10½	12	¼	337	15½	20	¼	152
10½	12	5/16	583	15½	20	5/16	267
11	12	.2	2:6	16	20	¼	148
11	12	¼	322	16	20	5/16	231
11	12	5/16	557				

See notes at foot of table about "Shells of Boilers," page 153.

Locomotives.

Baldwin Locomotive Works............................Philadelphia, Pa............See Page 78
Greene & Randolph...New York......................... " 54
H. N. Winans—(Boiler Powders)..................... " " .... " 108
Miller & Smith.. " " .... " 112
O. W. Child .... " " .... " 262
Porter, Bell & Co..Pittsburgh, Pa.............. " 128
Schenectady Locomotive Works......................Schenectady, N. Y......... " 148
Wiard Locomotive Attachment Co.—(Anti-Explosive)..New York......................See Page 110

Log Sawing Machine.

J. T. & R. H. Plass..New York......................See Page 208

Looking Glasses.

Leffingwell & Co.................................. ....Cleveland, Ohio........ ......See Page 190
See also, Glass—Looking.

Lubricating Oils.

Eclipse Lubric Oil Co.....................................New York.........See Page facing Title
Frazer Lubricator Co...................................... " " ....See Page 264
F. S. Pease...Buffalo, N. Y................. " 100
Isaac U. Coles...New York......................... " 20
P. H. Moffat..Chicago, Ill..................... " 190
Posts & Kalkman...New York......................... " 266
Wm. B. Browne & Co...... .... " " .... " 266

BOARD MEASURE—4 Pages.

Table gives Feet of Board Measure to One Running Foot of Board.

Area of a running foot multiplied by length in feet, will give superficial contents of board in square feet.

Application of Table:

A board is 1 inch thick, 1 foot 4¾ inches wide by 21 feet long. What is its contents? *Answer*, 1 foot equal 1, and 4¾ in. equal .3958, therefore, 1 foot 4¾ in. equal 1.3958, multiplied by 21, equal 29.3 feet, which is contents of board.

Thick. Inch.	Width of Board in Inches.								
	½	¾	1	1¼	1½	1¾	2	2¼	½
¼	.0104	.0156	.0208	.026	.0313	.0365	.0417	.0469	.0521
½	.0209	.0313	.0417	.0521	.0625	.0729	.0833	.0938	.1,42
¾	.0313	.0469	.0625	.0781	.0938	.1094	.125	.1406	.1563
1	.0417	.0625	.0833	.1042	.125	.1458	.1667	.1875	.2083
¼	.0521	.0781	.1042	.1302	.1563	.1823	.2083	.2344	.2604
½	.0625	.0938	.125	.1563	.1875	.2188	.25	.2813	.3125
¾	.0729	.1094	.1458	.1823	.2187	.2552	.2917	.3281	.3646
2	.0833	.125	.1667	.2083	.25	.2917	.3333	.375	.4166
¼	.0938	.1406	.1875	.2344	.2813	.3281	.375	.4219	.4688
½	.1042	.1563	.2083	.2604	.3125	.3646	.4167	.4688	.5208
¾	.1146	.1719	.2292	.2865	.3438	.401	.4583	.5156	.573
3	.125	.1875	.25	.3125	.375	.4375	.5	.5625	.625
¼	.1355	.2032	.2708	.3385	.4063	.4739	.5416	.6094	.677
½	.1457	.2186	.2917	.3646	.4375	.5104	.5833	.6563	.7292
¾	.1563	.2345	.3125	.3906	.4689	.5469	.625	.7031	.7813
4	.1666	.25	.3333	.4167	.5	.5833	.6667	.75	.8333
¼	.177	.2655	.3542	.4427	.5312	6198	.7083	.797	.8854
½	.1875	.2813	.375	.4688	.5625	.6563	.75	.8438	.9375
¾	.198	.297	.3958	.4948	.5938	.6927	.7917	.8905	.9895
5	.2083	.3125	.4167	.5208	.625	.7292	.8333	.9375	1.004
¼	.2187	.3231	.4375	.547	.6562	.7655	.875	.9845	1.095
½	.2291	.3437	.4583	.5729	.6875	.802	.9166	1.03	1.145
¾	.2395	.3594	.4791	.599	.7187	.8385	.9583	1.077	1.198
6	.25	.375	.5	.625	.75	.875	1.	1.125	1.25

Lubricators.

A. Carr..........New York..........See Page 244
Crane Bros. Manufacturing Co..........Chicago, Ill..........〃 230
Frazer Lubricator Co..........New York..........〃 264
Isaac U. Coles..........〃 〃 ....〃 20
J. G. Knapp Manufacturing Co..........〃 〃 ....〃 16
McNab & Harlin Manufacturing Co..........〃 〃 ....〃 220
Pancoast & Maule..........Philadelphia, Pa..........〃 26
Posts & Kalkman..........New York..........〃 266
Wiard Locomotive Attachment Co..........〃 〃 ....〃 110

Lumber.

Joseph Churchyard..........Buffalo, N. Y..........See Page 160

Machine Bits.

Champlin & Rogers..........Chicago, Ill..........See Page 192

Machine Bolts.

Wm. H. Haskell & Co; H. B. Newhall, Agent..New York..........See Page 166
Providence Tool Co.; 〃 〃 ....〃 166

Machine Forgings.

Pittsburgh Forge & Iron Co..........Pittsburgh, Pa..........See Page 174

Machine Oils.

Isaac U. Coles..........New York..........See Page 20
Wm. B. Browne & Co..........〃 〃 ....〃 266

See also, Lubricating Oils, Oils, &c., &c.

Machine or Nut Taps.

R. L. Howard & Son..........Buffalo, N. Y..........See Page 40
The Worcester Machine Screw Co.; H. B. Newhall, Agent..........New York..........See Page 166

See also, Taps; Nut Tappers, &c.

BOARD MEASURE—Continued.

Feet of Board Measure to One Running Foot of Board

(See application at head of table.)

Width of Board in Inches.

Thick. Inch.	2¾	3	3¼	3½	3¾	4	4¼	4½	4¾
¼	.0573	.0625	.0678	.073	.078	.0833	.0885	.0937	.099
½	.1145	.125	.1355	.1457	.1562	.1668	.177	.1875	.198
¾	.172	.1875	.203	.2188	.2344	.25	.2656	.3813	.297
1	.2292	.25	.2708	.2917	.3125	.3334	.3542	.375	.3958
¼	.2865	.3125	.3385	.3645	.3906	.4167	.4426	.4687	.4948
½	.3438	.375	.4062	.4375	.4687	.5	.5314	.5625	.5938
¾	.401	.4375	.474	.5105	.5470	.5834	.6198	.6563	.6927
2	.4583	.5	.5417	.5834	.625	.6667	.7084	.75	.7917
¼	.5155	.5625	.6095	.6563	.703	.75	.797	.8437	.8905
½	.573	.625	.677	.7292	.7814	.8334	.8855	.9375	.9895
¾	.6302	.6875	.7448	.802	.8595	.9167	.974	1.03	1.09
3	.6875	.75	.8125	.875	.9375	1.	1.063	1.125	1.188
¼	.7448	.8125	.8802	.948	1.016	1.083	1.15	1.22	1.286
½	.802	.875	.948	1.021	1.095	1.168	1.24	1.314	1.386
¾	.8595	.9375	1.015	1.095	1.172	1.25	1.327	1.406	1.485
4	.9167	1.	1.802	1.168	1.25	1.333	1.415	1.5	1.583
¼	.974	1.062	1.15	1.24	1.328	1.418	1.505	1.595	1.682
½	1.03	1.125	1.22	1.313	1.405	1.5	1.593	1.689	1.782
¾	1.086	1.187	1.285	1.385	1.485	1.584	1.68	1.73	1.88
5	1.146	1.25	1.355	1.457	1.567	1.667	1.77	1.875	1.98
¼	1.203	1.313	1.421	1.53	1.645	1.75	1.858	1.97	2.077
½	1.26	1.375	1.49	1.603	1.722	1.834	1.946	2.063	2.178
¾	1.319	1.438	1.556	1.675	1.8	1.916	2.034	2.155	2.275
6	1.375	1.5	1.625	1.75	1.875	2.	2.125	2.25	2.375

Machines for Emery Wheels.

D. Brewer & Co. ... Philadelphia, Pa. ... See Page 246

Machinery.

Buffalo Machinery Agency	Buffalo, N. Y.	See Page 160
C. Henry Hall & Co	New York	" 4
J. B. Waring	" "	" 24
J. S. Mundy	Newark, N. J	" 266
L. W. Pond	Worcester Mass	" 158
Miller & Smith	New York	" 112
S. S. Townsend	" "	" 2
Skinner & Gifford Manufacturing Co	Dunkirk, N. Y	" 140
Rand & Waring Drill & Compressor Co	New York	See Front Page
The Pratt & Whitney Co	Hartford, Conn	See Page 142
The Atlas Works	Pittsburgh, Pa	" 172
Wiard Locomotive Attachment Co.—(Anti-Explosive)	New York	" 110
Winans & Co	" "	" 106

Machine Screws.

Providence Tool Co.; H. B. Newhall, Agent	New York	See Page 166
The Worcester Machine Screw Co.; H. B. Newhall, Agent	" "	" 166

Machinists.

Bolen, Crane & Co	Newark, N. J	See Page 62
Snyder Brothers	Williamsport, Pa	" 46

Machinists' Supplies.

Champlin & Rogers	Chicago, Ill	See Page 192
Geo. H. Crain & Co	" "	" 124
Pancoast & Maule	Philadelphia, Pa	" 26
Posts & Kalkman	New York	" 263
Wm. B. Browne & Co	" "	" 266

See also, Machinists' Tools, &c.

BOARD MEASURE—CONTINUED

Feet of Board Measure to One Running Foot of Board.

(See application at head of table.)

Thick. Inch	Width of Board in Inches. 5	5¼	5½	5¾	6	6½	6¾	7	7½
¼	.1042	.1095	.1145	.1198	.125	.1355	.1405	.1458	.1563
½	.20 4	.2189	.2292	.2395	.25	.2708	.2813	.2917	.3125
¾	.3125	.328	.3438	.3597	.375	.4063	.422	.4375	.4687
1	.4166	.4375	.4584	.4792	.5	.5416	.5625	.5834	625
¼	.5208	.547	.573	.5 9	.625	.677	.703	.7292	.7814
½	.625	.6562	.6875	.7187	.75	.8125	.8438	.875	.9375
¾	.7292	.7655	.802	.8385	.875	.948	.9844	1.02	1.095
2	.8334	.875	.9166	.9584	1.	1.0·3	1 125	1.166	1.25
¼	.9375	.9845	1.03	1.077	1.125	1.22	1.266	1.314	1.406
½	1.042	1.095	1.145	1.198	1.25	1.354	1.406	1.459	1.563
¾	1.145	1.203	1 26	1.318	1.375	1.49	1.547	1.604	1 72
3	1.25	1.313	1.375	1.438	1.5	1.625	1.688	1.75	1.875
¼	1.354	1.422	1.4[illegible]	1.557	1.625	1.76	1.828	1.895	2.03
½	1.458	1.53	1.605	1.678	1.75	1.896	1.97	2.042	2.187
¾	1.564	1.64	1.72	1.797	1.875	2.031	2.109	2.187	2.345
4	1.668	1.75	1.834	1.917	2.	2.167	2.25	2.334	2.5
¼	1.77	1.857	1.918	2.335	2.125	2.302	2.39	2.48	2.655
½	1.875	1.97	2.063	2.155	2.250	2.438	2.53	2.625	2.813
¾	1 98	2.079	2.176	2.275	2.375	2 573	2.672	2.77	2 [illegible] 7
5	2.084	2.187	2.292	2.395	2.5	2.708	2.813	2.917	3.125
¼	2.187	2 297	2.405	2.515	2 625	2.845	2.954	3.063	3.28
½	2.2 3	2.406	2 52	2.634	2.750	2.98	3.095	3.209	3 437
¾	2.397	2.516	2.635	2.755	2.875	3.116	3.234	3.355	3.595
6	2.5	2.625	2.75	2.875	3.	3.25	3.375	3.5	3.75

Machinists' Tools.

Bolen, Crane & Co. ... Newark, N. J. ... See Page 62
Buffalo Machinery Agency ... Buffalo, N. Y. ... " 160
Geo. Worthington & Co. ... Cleveland, Ohio ... " 100
Isaac H. Shearman ... Philadelphia, Pa. ... " 80
L. W. Pond ... Worcester, Mass. ... " 158
Nelson Tool Works ... New York ... " 30
The Pratt & Whitney Co. ... Hartford, Conn. ... " 142
Thorne, De Haven & Co. ... Philadelphia, Pa. ... " 38
Wm. B. Bement & Son ... " " ... " 168
Wm. Sellers & Co. ... " " ... " 64

Malleable Iron Castings and Fittings.

Crane Bros. Manufacturing Co. ... Chicago, Ill. ... See Page 230
Phelps & Sanger ... Cleveland, Ohio ... " 120

Manufacturers Agents.

Winans & Co. ... New York ... See Page 106

Marine Forgings.

De Laney & Co. ... Buffalo, N. Y. ... See Page 98

Marking Cards.

W. W. Wilcox ... Chicago, Ill. ... See Page 190

Masons' Tools.

England & Bindley ... Pittsburgh, Pa. ... See Page 162

Mathematical Instruments.

F. Eckel ... New York ... See Page 76
F. S. Pease ... Buffalo, N. Y. ... " 100
Kuebler & Seelhorst ... Philadelphia, Pa. ... " 216
Wm. J. Young & Sons ... " " ... " 138

Mattocks.

Hermann Boker & Co. ... New York ... See Pages 152, 154

Mauls.

Hermann Boker & Co. ... New York ... See Pages 152, 154
Nelson Tool Works ... " " ... See Page 30

BOARD MEASURE—CONTINUED.

Feet of Board Measure to One Running Foot of Board.

(See application at head of table.)

Thick. Inch.	Width of Board in Inches.								
	8	8½	9	9½	10	10½	11	11½	12
¼	.1666	.1771	.1875	.198	.2083	.2187	.2292	.2395	.25
½	.3334	.3542	.375	.3958	.4167	.4375	.4583	.4792	.5
¾	.5	.5312	.5625	.5938	.625	.6562	.6875	.7187	.75
1	.6667	.7084	.75	.7916	.8334	.875	.9167	.9584	1.
¼	.8334	.8855	.9375	.9896	1.042	1.094	1.146	1.198	1.25
½	1.	1.062	1.125	1.187	1.25	1 312	1.375	1.437	1.5
¾	1.166	1.24	1.312	1.385	1.458	1.531	1.605	1.677	1.75
2	1.333	1.417	1.5	1.583	1.667	1.75	1.833	1.917	2.
¼	1.5	1.595	1.687	1.78	1.875	1.963	2.062	2.155	2.25
½	1.667	1.77	1.875	1.98	2.083	2.187	2.292	2.395	2.5
¾	1.834	1.948	2.063	2.177	2.292	2.406	2.521	2.635	2.75
3	2.	2.125	2.25	2.375	2.5	2.625	2.75	2.875	3.
¼	2.167	2.302	2.437	2.563	2.708	2.845	2.98	3.115	3.25
½	2.334	2.48	2.625	2.77	2.917	3.063	3.208	3.354	3.5
¾	2.5	2.655	2.812	2.97	3.125	3.28	3.437	3.594	3 75
4	2.667	2.834	3.	3.167	3 334	3.5	3.667	3.833	4.
¼	2.833	3.01	3.188	3.365	3.542	3.719	3.896	4.073	4.25
½	3.	3.187	3.375	3.562	3.75	3.938	4.125	4.313	4.50
¾	3.166	3.366	3.563	3.76	3.958	4.155	4.354	4.552	4.75
5	3.334	3.542	3.75	3 958	4 16	4.375	4.583	4.79	5.
¼	3.5	3.72	3.937	4.155	4 375	4.595	4.813	5.032	5.25
½	3.666	3.896	4.125	4.353	4 584	4.812	5.041	5.27	5.5
¾	3.834	4.072	4.313	4.553	4.793	5.03	5.271	5.51	5.75
6	4.	4.25	4.5	4.75	5.	5.25	5.5	5.75	6.

Measuring Tapes.

F. Eckel......New York...... See Page 76

Mechanics' Tools.

Newcomb Bro's Sons......New York...... See Page 88

Medical Apparatus—Electrical and Gutta Percha.

Bishop Gutta Percha Works......New York...... See Page 34
Charles T. Chester...... " " ...... " 212
Leclanche Battery Co...... " " ...... " 96

TO FIND SOLIDITY OF ROUND TIMBER.

When all the dimensions are in feet.

Length multiplied by square of ¼ mean girth equal cubic feet.

When length in feet, girth in inches.

Multiply as above, and divide by 144.

When all dimensions are in inches

Multiply as above, and divide by 1728.

TO FIND SOLIDITY OF SQUARE TIMBER.

When all dimensions are in feet.

Length multiplied by breadth multiplied by depth, equal cubic feet.

When either dimensions are in inches.

Multiply as above, and divide by 12.

When any two of the dimensions are in inches.

Multiply as above, and divide by 144.

Or use the following table as shown at foot of same.

TABLE ¼ GIRTHS.

¼ Girth. in inches.	Area in Feet.	¼ Girth in Inches.	Area in Feet.	¼ Girth in Inches	Area in Feet.
6	.250	12¼	1.04	19	2.50
¼	.272	½	1.08	½	2.64
½	.294	¾	1.12	20	2.77
¾	.317	13	1.17	½	2.91
7	.340	¼	1.20	21	3.06
¼	.364	½	1.26	½	3.20
½	.390	¾	1.31	22	3.36
¾	.417	14	1.36	½	3.51
8	.444	¼	1.41	23	3.67
¼	.472	½	1.46	½	3.83
½	.501	¾	1.51	24	4.00
¾	.531	15	1.56	½	4.16
9	.562	¼	1.61	25	4.34
¼	.594	½	1.66	½	4.51
½	.626	¾	1.72	26	4.69
¾	.659	16	1.77	½	4.87
10	.694	¼	1.83	27	5.06
¼	.730	½	1.89	½	5.25
½	.766	¾	1.94	28	5.44
¾	.303	17	2. 0	½	5.64
11	.840	¼	2.06	29	5.84
¼	.878	½	2.12	½	6.04
½	.918	¾	2.18	30	6.25
¾	.959	18	2.25		
12	1.000	½	2.37		

Area corresponding to the ¼ girth in inches multiplied by length in feet gives solidity in feet and decimal parts for either square or round timber.

In Round Timber take the *mean* girth. Square Timber take the side, which in practice corresponds with ¼ of the girth of Round Timber.

Merchant Iron.

Metal Blanks.

Metal Cornices.

Metallic Packing.

Metallic Paints.

Metallurgists.

Metals, Domestic and Imported.

WHEEL GEARING.—4 PAGES.

TERMS OR NAMES. (*Haswell.*)

The *Pitch Line* of a wheel, is the circle upon which the pitch is measured, and it is the circumference by which the diameter, or the velocity of the wheel, is measured.

The *Pitch* is the arc of the circle of the pitch line, and is determined by the number of the teeth in the wheel.

The *True Pitch (Chordial)*, or that by which the dimensions of the tooth of a wheel are alone determined, is a straight line drawn from the centres of two contiguous teeth upon the pitch line.

The *Line of Centres* is the line between the centres of two wheels.

The *Radius* of a wheel is the semi-diameter running to the periphery of a tooth.

The *Pitch Radius* is the semi-diameter running to the pitch line.

A *Mortice Wheel*, is a wheel constructed for the reception of teeth or cogs, which are fitted into recesses or sockets upon the face of the wheel

A wheel which impels another is termed the *Spur*, *Driver* or *Leader;* the on impell d is the *Pinion*, *Driven* or *Follower*.

A s ries of wheels in conn ction with each other is termed a *Train*.

When two wheels act upon one another, the greater is termed the *Wheel*, an the lesser the *Pinion*.

When a *Pinion is driven by a wheel*, the number of teeth in the pinion should not be less than eight.

When a *Wheel is driven by a pinion*, the number of teeth in a pinion should not be less than ten.

The *Number of teeth* in the wheel should not be divisible by the number of teeth in the pinion without a remainder This is in order to prevent the same teeth coming together so often as to cause an irregular wear of their faces.

PROPORTION OF TEETH OF WHEELS. (*Moleswarth.*)

From pitch line to top of tooth	= Pitch × 0.33
Total depth of teeth	= Pitch × 0.75
Thickness of tooth on pitch line	= Pitch × 0.45
Space between teeth on pitch line	= Pitch × 0.55
Thickness of rim of wheel	= Pitch × 0.45
Thickness of arms if flat	= Pitch × 0.45
Ordinary width of teeth in small pitches	= Pitch × 2.
" " " large "	= Pitch × 3.
Thickn ss round centre	= Pitch × 1.3

Mortice wheels to be wider than iron wheels by twice the thickness of the rim or by pitch × 0.9; their rim to be double the thickness of that of iron wheels.

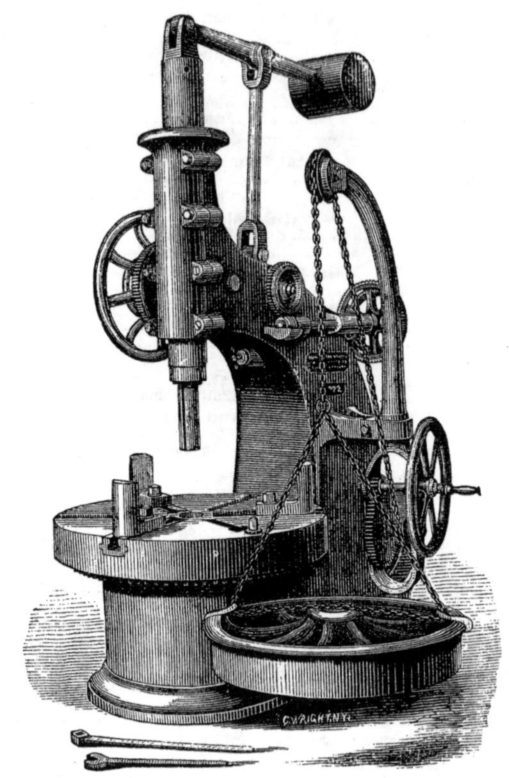

WHEEL GEARING—CONTINUED.

(Haswell.)

To compute the Number of Teeth in a Pinion or Follower to have a given Velocity.

RULE.—Multiply the velocity of the driver by its number of teeth, and divide the product by the velocity of the driven.

Example.—The velocity of a driver is 16 revolutions, the number of its teeth 54, and the velocity of the pinion is 48; what is the number of its teeth?

$16 \times 54 \div 48 = 18$ *teeth.*

2. A wheel having 75 teeth is making 16 revolutions per minute; what is the number of teeth required in the pinion to make 24 revolutions in the same time?

$16 \times 75 \div 24 = 50$ *teeth.*

To Compute the Diameter of a Wheel for a given Pitch and Number of Teeth.

RULE.—Multiply the diameter in the following table for the number of teeth by the pitch, and the product will give the diameter at the pitch circle.

Example.—What is the diameter of a wheel to contain 48 teeth of 2.5 ins. pitch?

$15.29 \times 2.5 = 38.225$ *ins.*

To Compute the Pitch of a Wheel for a given Diameter and number of Teeth.

RULE.—Divide the diameter of the wheel by the diameter in the table (page 171) for the number of teeth, and the product will give the pitch.

Example.—What is the pitch of a wheel when the diameter of it is 50.94 inches, and the number of its teeth 80?

$50.94 \div 25.47 = 2$ *ins.*

To Compute the Number of Teeth of a Wheel for a given Diameter and Pitch.

RULE.—Divide the diameter by the pitch, and opposite to the quotient in the table (page 171) is given the number of teeth.

Nail Rods.

England & Bindley....................................Pittsburgh, Pa...............See Page 162

Nails—Cut.

Collier & Scranton; Oxford Iron Co..............New York......................See Page 122
England & Bindley..Pittsburg, Pa.................. " 162
Geo. Worthington & Co...................................Cleveland, Ohio.............. " 100
Jones & Laughlins...... ....Pittsburg, Pa.................. " 32
John W. Quincy...New York .... " 178
Pratt & Co...Buffalo, N. Y................. " 132
Sidney Shepard & Co...................................... " .... " 196
Wm. Green & Co..Wilmington, Del............ " 248

See also, Nails—Wrought.

Nails—Wrought.

D. Brewer & Co...Philadelphia, Pa.............See Page 246
Jno. W. Quincy..New York....................... " 178
Pratt & Co...Buffalo, N. Y................... " 132
Wm. Green & Co ....Wilmington, Del............. " 248

See also, Nails—Cut.

WHEEL GEARING—Continued.

Pitch of Wheels. (*Haswell.*)

Showing the Diameter of a Wheel for a given Pitch, or the Pitch for a given Diameter.

No. of Teeth.	Diameter. Inches.	No. of Teeth.	Diameter. Inches.	No. of Teeth.	Diameter. Inches.	No. of Teeth.	Diameter. Inches.
8	2.61	45	14 33	82	26.11	119	37.88
9	2.93	46	14.65	83	26.43	120	38.2
10	3.24	47	14.97	84	26.74	121	38.52
11	3.55	48	15.29	85	27.06	122	38.84
12	3.86	49	15.61	86	27.38	123	39.16
13	4.18	50	15.93	87	27.7	124	39.47
14	4.49	51	16.24	88	28.03	125	39.79
15	4.81	52	16.56	89	28.33	126	40.11
16	5.12	53	16.88	90	28.65	127	40.43
17	5.44	54	17.2	91	28.97	128	40.75
18	5.76	55	17.52	92	29 29	129	41.07
19	6.07	56	17.8	93	29.61	130	41.38
20	6.39	57	18.15	94	29.93	131	41.7
21	6.71	58	18.47	95	30.24	132	42.02
22	7.03	59	18.79	96	30.56	133	42.35
23	7.34	60	19.11	97	30.88	134	42.66
24	7.66	61	19 42	98	31.2	135	42.98
25	7.98	62	19.74	99	31.52	136	43.29
26	8.3	63	20.06	100	31.84	137	43,61
27	8.61	64	20.38	101	32.15	138	43.93
28	8.93	65	20.7	102	32.47	139	44.25
29	9.25	66	21.02	103	32.79	140	44.57
30	9.57	67	21.33	104	33.11	141	44.88
31	9.88	68	21.65	105	33.43	142	45.2
32	10.2	69	21.97	106	33.74	143	45.52
33	10.52	70	22.29	107	34.06	144	45.84
34	10.84	71	22.61	108	34.38	145	46.16
35	11.16	72	22.92	109	34.7	146	46.48
36	11.47	73	23.24	110	35.02	147	46.79
37	11.79	74	23.56	111	35.34	148	47.11
38	12.11	75	23.88	112	35.65	149	47.43
39	12.43	76	24.2	113	35.97	150	47.75
40	12.74	77	24.52	114	36.29	151	48 07
41	13.06	78	24.83	115	36.61	152	48.39
42	13.38	79	25.15	116	36.93	153	48.7
43	13.7	80	25.47	117	37.25	154	49.02
44	14.02	81	25.79	118	37.56	155	49.35

Note. The pitch in this table is the *true pitch*, as before described.

Nickle.

Lucius Hart & Co	New York	See Page 56

Nuts, &c.

Collier & Scranton; Oxford Iron Co	New York	See Page 122
G. B. Walbridge	" "	" 130
Geo. Worthington & Co	Cleveland, Ohio	" 100
Greene & Randolph	New York	" 54
Hoopes & Townsend	Philadelphia, Pa	" 28
Lewis Oliver & Phillips, H. B. Newhall, Agent,	New York.	" 166
Miller & Smith	" "	" 112
New Haven Nut Co	New Haven, Conn	" 142
Lock, Nut & Bolt Co. of New York	New York	" 272
O. W. Child	" "	" 262
Phillipsburg Manufacturing Co	" "	" 22
Pittsburgh Bolt Co	Pittsburgh, Pa	" 118
Pratt & Co	Buffalo, N. Y	" 132
Providence Tool Co., H. B. Newhall, Agent	New York	" 166
Reading Bolt & Nut Works, H. B. Newhall, Agent	" "	" 166
Rhode Island Nut Co., H. B. Newhall, Agent	New York	See Page 166
T. B. Bickerton & Co	Philadelphia, Pa	" 138
Vose, Dinsmore & Co	New York	" 60
Wm. Gilmor of Wm	Baltimore, Md	" 248
Wm. H. Haskell & Co., H. B. Newhall, Agent	New York	" 166
Wm. Green & Co	Wilmington, Del	" 248

WHEEL GEARING—CONTINUED.

TEETH OF WHEELS—CAST-IRON. (*Molesworth.*)

Table showing the Horse-Power that may be transmitted by each Inch of Breadth of Tooth, with different Velocities and Pitches.

Veloc. in ft. per sec.	Pitch of Teeth in Inches.										
	¾	1	1¼	1½	1¾	2	2½	3	4	5	6
	h.p.	h.p.	h.p.	h.p.	h.p.	h.p.	h.p.	h.p.	h.p.	h.p.	h.p.
¼	.008	.015	.023	.033	.045	.06	.093	.135	.24	.37	.54
½	.017	.03	.047	.67	.09	.12	.18	.27	.43	.75	.108
¾	.025	.045	.07	.101	.138	.18	.281	.4	.72	1.12	1.62
1	.033	.06	.094	.135	.184	.24	.375	.54	.96	1.5	2.16
2	.067	.12	.188	.27	.366	.48	.75	1.08	1.9	3.0	4.3
3	.10	.18	.28	.40	.55	.72	1.1	1.6	2.8	4.5	6.4
4	.13	.24	.37	.54	.73	.96	1.5	2.1	3.8	6.	8.6
5	.17	.30	.47	.67	.91	1.2	1.8	2.7	4.8	7.5	10.8
6	.20	.36	.56	.81	1.1	1.4	2.2	3.2	5.7	9.	12.9
7	.23	.42	.65	.94	1.28	1.68	2.6	3.7	6.7	10.5	15.1
8	.27	.48	.75	1.1	1.4	1.9	3.	4.3	7.6	12.	17.2
9	.30	.54	.84	1.2	1.6	2.1	3.3	4.8	8.6	13.5	19.4
10	.33	.6	.94	1.35	1.8	2.4	3.7	5.4	9.6	15.	21.6
12	.40	.72	1.1	1.6	2.1	2.8	4.5	6.4	11.5	18.	25.9
14	.47	.84	1.3	1.8	2.5	3.3	5.2	7.5	13.4	21.	30.2
16	.54	.96	1.5	2.1	2.9	3.8	6.	8.6	15.3	24.	34.5
18	.61	1.1	1.7	2.4	3.3	4.3	6.7	9.7	17.3	27.	38.9
20	.66	1.2	1.9	2.7	3.6	4.8	7.5	10.8	19.2	30.	43.2
22	.74	1.3	2.1	2.9	4.	5.3	8.2	11.9	21.1	33.	47.5
24	.81	1.4	2.2	3.2	4.4	5.7	9.	12.9	23.	36.	51.8
26	.88	1.5	2.4	3.5	4.7	6.2	9.7	14.	24.9	39.	56.1
28	.95	1.6	2.6	3.7	5.1	6.7	10.5	15.1	26.9	42.	60.4
30	1.01	1.8	2.8	4.	5.5	7.2	11.2	16.2	28.8	45.	64.8
35	1.2	2.1	3.3	4.7	6.4	8.4	13.1	18.9	33.6	52.5	75.6
40	1.3	2.4	3.7	5.4	7.3	9.6	15.	21.6	38.4	60.	86.4

PITCHES OF EQUIVALENT STRENGTH FOR THE TEETH OF WHEELS IN DIFFEREMT MATERIALS.

Pitch for cast-iron = 1.00
" brass = 1.12
" hard wood = 1.26

Nitro Glycerine.

Geo. W. Mowbray........North Adams, Mass.........See Page 182

Nut Locks.

Jno. W. Quincy........New York........See Page 178
Lock, Nut & Bolt Co. of New York........ " " .... " 272
Vose, Dinsmore & Co........ " " .... " 60

Nuts, Hot Pressed.

New Haven Nut Co........New Haven, Conn..........See Page 142

WHEELS—SIZES AND WEIGHT.

Double Plate Wheels.

W. W. Snow's Patent, for Inside or Outside Bearings.

36 in.	diameter	Extra heavy, for Engine Trucks	640 lbs.
33 in.	"	" " " "	600 "
33 in.	"	Engines, Tenders, and Passenger Cars	530 "
30 in.	"	Extra heavy, for Engine Trucks	520 "
30 in.	"	Engines, Tenders, and Cars	4[illegible]0 "
28 in.	"	" " "	440 "
26 in.	"	" " "	380 "
24 in.	"	" " "	320 "
20 in.	"	" " "	280 "

Atwood-Washburn Patent Wheels.

36 in.	diameter	Extra heavy, for Engine Trucks	62 lbs.
33 in.	"	" " " "	[illegible]80 "
33 in.	"	Engine, Tender, and Passenger Cars	[illegible]30 "
33 in.	"	Ordinary and general use	5[illegible]0 "
33 in.	"	Light Freight Cars	480 "
30 in.	"	Extra heavy, for Engine Trucks	510 "
30 in.	"	Engines, Tenders, and Cars	48[illegible] "
30 in.	"	Light Freight Cars	45[illegible] "
30 in.	"	Coal Cars	430 "
28 in.	"	Engine Trucks and Tenders	430 "
26 in.	"	" " "	380 "
26 in.	"	Coal Cars, heavy	365 "
26 in.	"	" light	345 "
24 in.	"	Engine Trucks and Tenders	330 "
20 in.	"	" " "	280 "

Single Plate Wheels.

33 in.,	Single Plate	Freight Cars, heavy	500 lbs.
33 in.,	"	" light	480 "
30 in.,	"	Engine Trucks and Tenders	46[illegible] "
30 in.,	"	Freight and Coal Cars	440 "
30 in.,	"	" " "	420 "
28 in.,	"	Engine Trucks and Tenders	420 "
26 in.,	"	" " "	375 "
26 in.,	"	Coal Cars	330 "
24 in.,	"	Engine and Tender Trucks	33[illegible] "
20 in.,	"	" " "	280 "

Engine and Tender Truck Wheels.

With Hollow Spokes and Rim.

36 in.	diameter	600 lbs.
33 in.	"	530 "
30 in.	"	480 "
28 in.	"	430 "
26 in.	"	380 "

Construction Wheels.

24 in.	diameter,	3 in.	tread,	3/4 in.	flange	200 lbs
24 in.	"	2 in.	"	3/4 in.	"	140 "
20 in.	"	5 in.	"	1 1/8 in.	"	160 "
20 in.	spoke,	5 in.	"	1 1/2 in.	"	100 "
20 in.	"	2 1/2 in.	"	3/4 in.	"	120 "
16 in.	"	2 in.	"	3/4 in.	"	70 "

Nuts, Hot & Cold Pressed.

Providence Tool Co.; H. B. Newhall, Agent.....New York.................... See Page 166

Nut Tappers.

Isaac H. Shearman..................................Philadelphia, Pa...........See Page 80
Wm. B. Bement & Son.................................. " .... " 168

Oakum.

De Grauw, Aymar & Co.................................New York ....See Page 112

Office Chairs.

A. L. Hale & Bro...Chicago, Ill..................See Page 206

Office Furniture.

A. L. Hale & Bro...Chicago, Ill..................See Page 206
Joseph Churchyard.......................................Buffalo, N. Y................ " 160

Oil Cabinets.

D. Brewer & Co. ....Philadelphia, Pa...........See Page 246

Oil Cans and Cups.

J. G. Knapp Manufacturing Co.....................New York.......................See Page 16
Orrin L. Gridley..Buffalo, N. Y................. " 196
Pancoast & Maule...Philadelphia .... " 26
Parmélee & Bonnell.......................................Buffalo, N. Y................. " 202

Oil of Vitriol.

Van Tuyl Manufacturing Co.........................New York ....See Page 242

Oils.

Eclipse Lubric Oil Co ....New York.........See Page Facing Title
F. S. Pease...Buffalo, N. Y.................See Page 100
P. H. Moffat..Chicago, Ill.................... " 190
Wetherill & Bro...Philadelphia, Pa.. .... " 52
Wm. B. Browne & Co....................................New York........................ " 263

Oil Tank Cars.

Snyder Brothers...Williamsport, Pa............See Page 46

Oil Well Casing.

Evans, Dalzell & Co......................................Pittsburgh, Pa...............See Page 58
Morris, Tasker & Co......................................Philadelphia, Pa.......... " 36

Oil Well Tubing.

Evans, Dalzell & Co......................................Pittsburgh, Pa...............See Page 58
Morris, Tasker & Co......................................Philadelphia, Pa............ " 36

Old Rails.

Holmes & Lissberger.....................................New York ....See Page 84
John W. Quincy.. " .... " 178
Miller & Smith.. " .... " 112
O. W. Child .... " .... " 262

TESTS OF STEEL AND IRON.

(*Scientific American.*)

Nitric acid will produce a black spot on steel; the darker the spot the harder the steel. Iron, on the contrary, remains bright if touched with nitric acid.

Good steel in its soft state has a curved fracture and a uniform grey lustre; in its hard state a dull, silvery, uniform white. Cracks, threads, or sparkling particles denote bad quality.

Good steel will not bear a white heat without falling to pieces, and will crumble under the hammer at a bright red heat, while at a middling heat it may be drawn out under the hammer to a fine point.

Care should be taken that before attempting to draw it out to a point, the fracture is not concave, and should it be so the end should be filed to an obtuse point before operating. Steel should be drawn out to a fine point and plunged into colp water; the fractured point should scratch glass. To test its toughness, place a fragment on a block of cast iron; if good it may be driven by the blow of a hammer into the cast-iron, if poor it will crush under the blow.

A soft, tough iron, if broken gradually, gives long silky fibres of leaden-grey hue, which twist together and cohere before breaking.

A medium even grain with fibres denotes good iron.

Badly refined iron gives a short blackish fibre on fracture. A very fine grain denotes hard steely iron, likely to be cold-short and hard.

Coarse grain with bright crystallized fracture or discoloured spots denotes cold-short, brittle iron, which works easily when heated and welds well. Cracks on the edge of a bar are indications of hot-short iron. Good iron is readily heated, is soft under the hammer, and throws out few sparks.

BEAMS AND CHANNELS.

Sizes and Strength.

(Trenton Iron Works.)

Height in Inches.	Weight per Yd. in Pounds.	Width of Flange in Inches.	Thickness of Webb in Inches.	Co-efficient for Strength.
BEAMS.				
15 1/8	200	5 3/4	.6	748 000
15 3/16	150	5	1/2	551 000
12 5/16	170	5 1/2	.6	511 000
12 1/4	125	4.79	.48	377 000
10 1/2	135	5	.47	360 (00
10 1/2	105	4 1/2	3/8	286 000
9	125	4 1/2	.58	268 000
9	85	4	3/8	189 000
9	70	3 1/2	.3	152 000
8	80	4 1/2	3/8	168 000
8	65	4	.3	135 (00
7	60	3 1/2	3/8	101 000
6	50	3 1/2	.3	76 800
6	40	3	1/4	62 600
5	40	3	5/16	49 100
5	30	2 3/4	1/4	38 700
4	37	3	5/16	36 800
4	30	2 3/4	1/4	30 100
CHANNELS.				
12 1/4	140	4	11/16	384 000
12 1/4	85	3	7/16	242 000
9	70	3 1/8	7/16	147 000
9	50	2 1/2	.34	101 000
6	45	2 1/2	3/8	55 400
6	33	2 1/4	.28	41 800

To find the safe load in pounds when the beam is supported at each end, and the load is uniformly distributed over the span.

RULE—Divide the number given for the beam in the column headed "Co-efficient for Strength," by the distance between supports estimated in feet. The load so found will be nearly one-third of the ultimate or breaking strength of the beam. When the load is concentrated entirely at the centre of the beam, one-half the above amount must be taken.

The Deflection in inches, at the middle of the span, for such distribnted load, will be found by dividing the square of the span, taken in feet, by seventy (70) times the depth of the beam, taken in inches.

Example—What uniformly distributed load will a 12¼ inch beam of 125 lbs. per yard, and having a clear span of 15 feet, bear with safety, and what will be the deflection under this load?

$$\text{Ans.}\ \frac{377000}{15} = 25{,}133 \text{ lbs.} = \text{Safe Load.}$$

$$\frac{15\text{x}15}{70\text{x}12\frac{1}{4}} = 0.26 \text{ in.} = \text{Deflection.}$$

As larger co-efficients for similar beams are given by some other makers, it is proper to add that such co-efficients are based upon a larger proportion of the ultimate strength than is sanctioned by any good authority for practical use, and not upon any difference in the section or quality of the beam.

Paints for Railroads.

C. E. Hecht.......................Easton, Pa.......................See Page 158
John W. Masury & Son.......................New York.......................“ 214
Prince's Metallic Paint Co.......................“ “ ....“ 236
See also, Paints.

Parallel Bench Vise.

R. L. Howard & Son.......................Buffalo, N. Y.......................See Page 40

Passenger Cars.

Greene & Randolph.......................New York.......................See Page 54
O. W. Child.......................“ “ ....“ 262
See also, Car Builders.

Patch Bolts for Boilers.

Pittsburgh Bolt Co.......................Pittsburgh, Pa.......................See Page 118

Patent Cold Rolled Shafting.

Geo. V. Cresson.......................Philadelphia, Pa.......................See Page 138
Jones & Laughlins.......................Pittsburgh, Pa.......................“ 32

Patent Couplings.

Geo. V. Cresson.......................Philadelphia, Pa.......................See Page 138

Paving Tools.

Nelson Tool Wooks.......................New York.......................See Page 30

Pens.

Joseph Gillott & Sons.......................New York.......................See Page 86

Permanent Wood Filling.

Valentine & Co.......................New York.......................See Page 198

CROWNS OF ARCHES.

Thickness required for Crowns of Arches.

(Hurst.)

Radius of Curvature.	Stone Arches.	Brick Arches.	Radius of Curvature.	Stone Arches.	Brick Arches.	Radius of Curvature.	Stone Arches.	Brick Arches.
Feet.	Feet.	Feet.	Feet.	Feet.	Feet	Feet.	Feet.	Feet.
2	.42	.56	14	1.12	1.50	60	2.33	3.10
2½	.47	.63	15	1.16	1.55	65	2.42	3.22
3	.52	.69	16	1.20	1.60	70	2.51	3.35
3½	.56	.75	17	1.24	1.65	75	2.60	3.46
4	.60	.80	18	1.27	1.70	80	2.68	3.58
4½	.64	.85	19	1.32	1.74	85	2.77	3.69
5	.67	.90	20	1.34	1.79	90	2.85	3.80
5½	.71	.94	22	1.41	1.88	95	2.92	3.90
6	.74	.98	24	1 47	1.96	100	3.00	4.00
7	.80	1.06	25	1.50	2.00	110	3.15	4.20
8	.85	1.13	30	1.64	2.19	120	3.29	4.38
9	.90	1.20	35	1.78	2.37	130	3.42	4.56
10	.95	1.26	40	1.90	2.53	140	3.35	4.73
11	1.00	1.33	45	2.01	2.68	150	3.67	4.90
12	1.04	1.38	50	2.12	2.83	160	3.80	5.06
13	1.08	1.44	55	2.22	2.97	170	4.18	5.58

PASTE THAT WILL KEEP A YEAR.

Dissolve a teaspoonful of alum in a quart of warm water. When cold, stir in as much flour as will make it the consistency of thick cream, being particular to beat up all the lumps; stir in as much powdered resin as will stand on a dime, and pour in a few drops of oil of cloves to give it a pleasant odor. Have on the fire a teacupful of boiling water; pour the flour mixture into it, stirring well all the time. In a few minutes it will be of the consistency of mush. Pour it into an earthen or china vessel; let it cool; lay a cover on, and put it in a cool place. When needed for use, take out a portion and soften it with warm water. Paste thus made will last twelve months. It is better than gum, as it does not gloss the paper, and can be written upon.

Picks.

England & Bindley	Pittsburgh, Pa	See Page 162
G. B. Walbridge	New York	" 130
Geo. Worthington & Co	Cleveland, Ohio	" 100
Hermann Boker & Co	New York	See Pages 152, 154
Nelson Tool Works	" "	See Page 30
Providence Tool Co.; H. B. Newhall, Agent	" "	" 166
T. B. Bickerton & Co	Philadelphia, Pa	" 138

Pig Iron.

Chas. W. Matthews	Philadelphia, Pa	See Page 38
Collier & Scranton; Oxford Iron Co	New York	" 122
E. L. Hedstrom	Buffalo, N. Y	" 202
Holmes & Lissberger	New York	" 84
John A. Griswold & Co	Troy, N. Y	" 50
Jonas S. Heartt & Co	" "	" 134
John W. Quincy	New York	" 178
Miller & Smith	" "	" 112
Posts & Kalkman	" "	" 266
Pratt & Co	Buffalo, N. Y	" 132
Rhodes & Bradley	Chicago, Ill	" 192
Rogers & Co	" "	" 144

INSTRUCTIONS FOR HANDLING AND USING MOWBRAY'S TRI-NITRO-GLYCERIN.

1. Handle carefully, avoiding a sudden jar or concussion, and be very careful, if any is spilt outside the can, to avoid striking it against any hard substance.

2. When solid, thaw out by placing the cans in a tub of warm water, not hotter than the wrist can bear, first pouring warm water into the can, and always remove the can before adding more hot water to the tub.

3. To fill Cartridges, &c.—Hold the cartridge to be filled over a tray, say two feet by three feet, the bottom of which should be covered with Plaster of Paris (which will not readily explode when saturated with Nitro-Glycerin.) The soiled Plaster of Paris should be frequently renewed.

4. If the Nitro-Glycerin in a liquid state is kept in store or magazine for some time, the cork should be loosely inserted, and a pint of cold water poured in each can, to be frequently poured off and replaced with fresh cold water in warm weather, taking care to retain the bladder under the cork. It is preferable, when ice can be procured, to congeal the Nitro-Glycerin.

5. Use Funnels (gutta-percha if it can be had) for filling water holes. Under no circumstances whatever attempt to tamp the drill holes; it is unnecessary, and may kill the man who attempts it.

6. Hot irons to warm the water, or soldering the cans, will be sure to cause explosions.

7. Never sledge or attempt drilling in a hole or seam where Nitro-Glycerin has been spilled; fire an exploder, which will effectually clear it up.

8. Never pour Nitro-Glycerin into a hole unless perfectly sure that it is a sound hole, or will hold water; if seamy always use cartridges.

9. To obtain the best results with Nitro-Glycerin, drill deep holes, 6 feet or more. Use powerful exploders and well insulated wires, It is cheaper to fire by electric battery with simultaneous explosion, than to fire several holes with tape fuse.

10. Look out after a blast for any unexploded cartridges lying around.

11. Never allow any but the most careful persons to handle or have charge of the Nitro-Glycerin, and insist upon the use of every precaution to prevent an accident or explosion.

12. Never allow empty Glycerin cans to be used for any other purpose, but destroy them by a fuse and exploder, or building a fire under them, first, however, removing them to a safe distance.

13. Examine your cans from time to time, and notice if, at the level of the Nitro-Glycerin, any pin-holes have eaten through; in such case procure a new can, or stone jar, and empty the contents out, not trusting your hold to the upper part of the can, lest it may give way.

14. When solid, or congealed, it is absolutely safe; if possible, therefore, any surplus should be stored surrounded with ice, since no explosion can take place whn it is solid.

THE ELECTRIC RAILROAD SIGNAL CO.

194 Fulton Street, New York,

Are now prepared to receive orders for the erection of their Improved and Perfected **ELECTRIC SIGNALS,** manufactured under the patents of **F. L. POPE** and **S. C. HENDRICKSON.** Our Improved

ELECTRIC SEMAPHORE

can be operated at any distance, and its indications are instantly and infallibly **REPEATED BACK** to the signalman on the same wire. For STATIONS, TUNNELS, BRIDGES, and SECTIONS OF SINGLE TRACK, this Signal is invaluable. We also furnish an

AUTOMATIC BLOCK SYSTEM,

which is operated by the trains themselves, a section of the track being brought into the electric circuit. This system is the **ONLY RELIABLE** one for working automatic signals, and is covered by the original patent of F. L. POPE, July 16, 1872. ☞ *Our Signals have been thorougly tested in practical service.* ☜ We also design and erect to order, special Signals for

Draw-Bridges, Crossings, Station-Yards, Etc., Etc.

J. N. ASHLEY,
Business Manager.

F. L. POPE,
Superintendent.

J. W. STOVER,
Agent for the New England States.

F. L. POPE & CO.

194 Fulton Street, New York,

MANUFACTURERS AND DEALERS IN

TELEGRAPH INSTRUMENTS,

APPARATUS AND SUPPLIES,

OF EVERY DESCRIPTION.

We invite the attention of PURCHASING AGENTS and RAILROAD TELEGRAPH SUPERINTENDENTS to our new patterns of

STANDARD TELEGRAPH INSTRUMENTS,

which have been designed **EXPRESSLY FOR RAILROAD SERVICE.** They are of novel and elegant design, superior finish, and especially constructed for **STRENGTH** and **DURABILITY.** ☞ Our magnets are all wound with tested copper wire of 95 PER CENT conductivity, and by a patented process which adds 20 per cent to their working strength.

☞ *Send for our Illustrated Catalogue and Price List of Instruments, Supplies and Material.* ☜

194 FULTON ST., NEW YORK. [P. O. Box 6010.]

Pig Lead.

E. W. Blatchford & Co.....Chicago, Ill.....See Page 124
Lucius Hart & Co.....New York....." 56

Pipe Tongs.

Barwick Wrench Co.....Boston, Mass.....See Page 60

WEIGHTS.

Avoirdupois Weight.

The Grain is the same in Troy, Apothecaries and Avoirdupois Weights.

The standard avoirdu o's pound is the weight of 27.7015 cubic inches of distilled water weighed in the air, at 35 85 degrees Fahr., barometer at 30 inches.

27.343 grains = 1 drachm.

drachms.	ozs.	lbs.	qrs.	cwts.	ton.	French grammes.
1 =	.0625 =	.0039 =	.000139 =	.000035 =	.00000174 =	1.771846
16 =	1 =	.0625 =	.00223 =	.000558 =	.000028 =	28.34954
256 =	16 =	1 =	.0357 =	.00893 =	.000447 =	453.59
7168 =	448 =	28 =	1 =	.25 =	.0125 =	12700
28672 =	1792 =	112 =	4 =	1 =	.05 =	50802
573440 =	35840 =	2240 =	80 =	20 =	1 =	1.016048

A stone = 14 pounds. A quintal = 100 pounds.

Troy Weight.

For Gold, Silver and Precious Metals.

grains.	dwts.	ozs.	lb.	French grammes.
1 =	.04167 =	.00208 =	.0001736 =	.0648
24 =	1 =	.05 =	.004167 =	1.555
480 =	20 =	1 =	.0833 =	31.1035
5760 =	240 =	12 =	1 =	373.242

175 lbs. Troy = 144 Avoirdupois.
lbs. Avoirdupois × .82286 = lbs. Troy.
lbs. Troy × 1.2153 = lbs. Avoirdupois.

The jeweler's Carat is equal, in the United States, to 3.2 grains; in London, to 3.17 grains; in Paris, to 3.18.

Pure Gold is worth	$20.67 per oz. Troy,	or	$18.84 per oz. Avoirdupois.
" Silver "	$1.36 "	"	$1.24 "	"
Standard Gold "	$18.60 "	"	$16.96 "	"
" Silver "	$1.225 "	"	$1.117 "	"

Apothecaries' Weight.

United States and British.

20 grains..... 1 scruple.
3 scruples..... 1 dram = 60 grs.
8 drams..... 1 ounce = 24 scruples = 480 grs.
12 ounces..... 1 pound = 96 drams = 288 scruples = 5760 grains.

In Troy and Apothecaries' weights, the grain, ounce, and pound are the same.

FRENCH WEIGHTS,

Equivalent in Avoirdupois.

	Grains.	Ounces.	Pounds.	Tons. 2240 lbs.
Milligramme	.015433	——	——	——
Centigramme	.1 43 1	.000352	.000022	——
Decigramme	1.54331	.003527	.000220	——
Gramme	15.4331	.035275	.002204	——
Decagramme	1 4.331	.352758	.022047	——
Hectogramme	1543.31	3.52758	.220473	.000 98
Kilogramme	15433.1	35.2758	2.20473	.000984
Myriogramme	——	352.758	22.0473	.009842
Quintal	——	3527.58	220.473	.098425
Millier or Tonne	——	35275.8	2204.73	.984258

Piston Rods.

Anderson & Woods....................Pittsburgh, Pa..............See Page 150
Jones & Laughlins.................... " .... " 32

Pit Car Wheels.

The Atlas Works....................Pittsburgh, Pa..............See Page 172

Pivot Bridges.

Wm. Sellers & Co....................Philadelphia, Pa..............See Page 64

Planes.

England & Bindley....................Pittsburgh, Pa..............See Page 162

Planing Machines.

Isaac H. Shearman....................Philadelphia, Pa..............See Page 80
L. W. Pond....................Worcester, Mass.............. " 158
The Pratt & Whitney Co....................Hartford, Conn.............. " 142
Wm. B. Bement & Co....................Philadelphia, Pa.............. " 168
Wm. Sellers & Co.................... " .... " 64

Planing Mills.

C. B. Rogers & Co....................New York..............See Page 210

MEASURES.—4 PAGES.

LONG MEASURE.

ins.	feet.	yards.	fath.	poles.	furl.	mile.	French mètres.
1 =	.083 =	.02778 =	.0139 =	.005 =	.000126 =	.0000158 =	.0254
12 =	1 =	.333 =	.1667 =	.0606 =	.00151 =	.0001894 =	.3048
36 =	3 =	1 =	.5 =	.182 =	.00454 =	.000568 =	.9144
72 =	6 =	2 =	1 =	.364 =	.0091 =	.001136 =	1.8287
198 =	16½ =	5½ =	2¾ =	1 =	.025 =	.003125 =	5.0291
7920 =	660 =	220 =	110 =	40 =	1 =	.125 =	201.16
63360 =	5280 =	1760 =	880 =	320 =	8 =	1 =	1609.315

A palm = 3 inches. A span = 9 inches.
A hand = 4 " A cable's length = 120 fathoms.

SURVEYING MEASURE (Lineal).

ins.	links.	feet.	yards.	chains.	mile.	French mètres.
1 =	.126 =	.0833 =	.0278 =	.00126 =	.0000158 =	.0254
7.92 =	1 =	.66 =	.22 =	.01 =	.000125 =	.2012
12 =	1.515 =	1 =	.333 =	.01515 =	.000189 =	.3048
36 =	4.545 =	3 =	1 =	.045 5 =	.000568 =	.9144
792 =	100 =	66 =	22 =	1 =	.0125 =	20.116
63360 =	8000 =	5280 =	1760 =	80 =	1 =	1609.315

1 knot or geographical mile = 6082.66 feet = 1854 mètres = 1.152 statute mile.
1 Admiralty knot = 1.1515 = 6080 feet.

FRENCH LONG MEASURE.

	Inches.	Feet.	Yards.	Miles.
Millimetre	.03986	.00328	——	——
Centimetre	.39368	.0328	——	——
Decimetre	3.9368	.32807	.109357	——
Metre	39.368	3.2807	1.09357	——
Decametre	393.68	32.807	10.9357	——
Hectometre	——	328.07	109.357	.0621347
Kilometre	——	3280.7	1093.57	.6213466
Myriametre	——	32807.	10935.7	6.213466

TIME.

Seconds.	Minutes.	Hours.	Days.	Weeks.
60 =	1			
3600 =	60 =	1		
86400 =	1440 =	24 =	1	
604800 =	10080 =	168 =	7 =	1

Sidereal Day = 23 h., 56 m., 4.092 sec., in solar or mean time.
Solar Day (mean) = 24 h., 3 m., 56.555 sec., in sideral time.
Sideral year, or revolution of the earth, 365.25635 solar days.
Solar or Calendar Year, 365.24224 solar days.

Plated Knobs.

Romer & Co.....Newark, N. J.....See Page 204

Plate Glass.

D. R. Hobart & Co.....New York.....See Page 126
Leffingwell & Co.....Cleveland, Ohio..... " 190

Pliers.

Biddle Manufacturing Co.....New York.....See Page 12

Plow Clevises.

Crane Bros. Manufacturing Co.....Chicago, Ill.....See Page 230

Plows—Railroad.

Nelson Tool Works.....New York.....See Page 30

Plumbers' Castings.

Philadelphia Smelting Co.....Philadelphia, Pa.....See Page 134

Plumbers' Supplies.

A. Carr.....New York.....See Page 244
A. Fulton's Son & Co.....Pittsburgh, Pa..... " 116
McNab & Harlin Manufacturing Co.....New York..... " 220

Plumbers' Tools.

Phelps & Sanger.....Cleveland, Ohio.....See Page 120

Plush Buttons.

L. G. Tillotson & Co.....New York.....See Page 94

Pop Valves.

Posts & Kalkman.....New York.....See Page 266

Porcelain Door Knobs.

Romer & Co.....Newark, N. J.....See Page 204

MEASURES.—Continued.

Square Measure.

ins.		feet.		yards.		perches.		roods.		acre		square mètres.
1	=	.00694	=	.000772	=	.0000255	=	.00000064	=	.000000159	=	.000645
144	=	1	=	.111	=	.00367	=	.0000918	=	.000023	=	.0929
1296	=	9	=	1	=	.0331	=	.000826	=	.0002062	=	.8361
39204	=	272¼	=	30¼	=	1	=	.025	=	.00625	=	25.292
1568160	=	10890	=	1210	=	40	=	1	=	.25	=	1011.7
6272640	=	43560	=	4840	=	160	=	4	=	1	=	4046.7

100 square feet = 1 square.
1 chain wide = 8 acres per mile.
10 square chains = 1 acre.
1 hectare = 2.471143 acres.
1 square mile { = 27878400 sq. feet. = 3097600 sq. yards. = 640 acres.
Acres × .0015625 = square miles.
Sq. yds. × .000000323 = sq. miles.
A section of land is 1 mile square, and contains 640 acres.
A square acre is 208.71 feet at each side.
" ½ " 147.58 " "
" ¼ " 104.355 " "
A circular " 235.504 feet in diameter.
" ½ " 166.527 " "
" ¼ " 117.752 " "

FRENCH SQUARE MEASURE.

	Sq. Ins.		Sq. Feet.		Sq. Yds.		Acres.
Sq. Millimetre	.00154	...	.0000107	...	.000001	...	——
Sq. Centimetre.	.15498	...	.0010763	...	.000119	...	——
Sq. Decimetre.	15.498	...	.1076305	...	.011958	...	——
Sq. Metre, or Centiare...	1549.8	...	10.76305	...	1.19589	...	.000247
Sq. Decametre of Are...	15498.	...	1076.305	...	119.589	...	.024709
Hectare	——	...	107630.5	...	11958.9	...	2.4706
Sq. Kilometre.	.38607 sq. miles	...	1076305.	...	119589.	...	247.086
Sq. Myriametre	38.607 sq. miles	...	——	...	——	...	24708.6

Portable Engines.

Isaac H. Shearman........Philadelphia, Pa........See Page 80
J. S. Mundy........Newark, N. J........ " 266
Rand & Waring Drill & Compressor Co........New York........See Front Page
Wm. P. Kellogg & Co........Troy, N. Y........See Page 98

Portable Forges.

Newcomb Bros. Sons........New York........See Page 88
S. S Townsend........ " " .... " 2
T. B. Bickerton & Co........Philadelphia, Pa........ " 138
W. P. Kellogg & Co........Troy, N. Y........ " 98

Powder.

Laflin & Rand Powder Co........New York........See Page next to Back Cover

Power Machines.

Biddle Manufacturing Co........New York........See Page 12

Presses—Copying.

T. Shriver & Co........New York........See Page 194

Printers.

Allen, Lane & Scott........Philadelphia, Pa........See Page 90
Chas. F. Ketcham........New York........ " 262
Warren, Johnson & Co........Buffalo, N. Y........ " 100

Pulleys.

Bolen Crane & Co........Newark, N. J........See Page 62
Buffalo Machinery Agency........Buffalo, N. Y........ " 160
Erie City Iron Works........Erie, Pa........ " 170
Geo. V. Cresson........Philadelphia, Pa........ " 138
R. L. Howard & Son........Buffalo, N. Y........ " 40

MEASURES.—Continued.

Cubic Measure.

ins.	feet.	yard.	cubic metrès.
1 =	.0005788 =	.000002144 =	.000016386
1728 =	1 =	.03704 =	.028315
46656 =	27 =	1 =	.764513.

A cord of wood = 128 cubic feet, being 4 feet high, 4 feet wide, and 8 feet long.
42 cubic feet = a ton of shipping.

A Cubic Foot is Equal to

1728 cubic inches.
.037037 cubic yard.
.803564 U. S. struck bushel of 2150.42 cubic inches.
3.21426 U. S pecks.
7.48052 U. S liquid galls. of 231 cub. inch.
6.42851 U. S. dry galls.
29.92208 U. S. liquid quarts.
25.71405 U. S. dry quarts
59.84416 U. S. liquid pints.
51.42809 U. S. dry pints.
239.37662 U. S. gills.
.26667 flour barrel of 3 struck bushels.
.23748 U. S. liquid barrel of 31½ galls.

FRENCH CUBIC OR SOLID MEASURE.

	Gill	Pint.	Quart	Gall.	Peck.	Bushel.	Cubic Inches.	Cubic Feet.
Centilitre, Dry	—	.0181	—	—	—	—	.61 16	—
Liquid	.0845	.0211	—	—	—	—		
Decilitre, Dry	—	.1816	.0908	—	.0113	—	6.1016	—
Liquid	.8452	.2113	.1056	.0264	—	—		
Litre, Dry	—	1.816	.908	—	.1135	—	61.016	.0353
Liquid	8.452	2.113	1.056	.2641	—	—		
Decalitre, Dry	—	—	9.08	—	1.135	.2837	610.165	.3531
Liquid	84.52	21.13	10.56	2.641	—	—		
Hectolitre, Dry	—	—	90.8	—	11.35	2.837	6101.6	3.531
Liquid	—	211.3	105.6	26.41	—	—		
Kilolitre or Cubic Meter, Dry	—	—	—	—	113.5	28.37	61016.	35.31
Liquid	—	—	1056.5	264.1	—	—		
Myriolitre, Dry	—	—	—	—	1135.	283.72	—	353.1
Liquid	—	—	10565.	2641.4	—	—		

CHAMPLIN & ROGERS,

No. 155 Fifth, Avenue, CHICAGO, ILL.,

DEALERS IN

Taps, Dies, Reamers, Screws, Screw Plates, Drills, Gas Fitters' Tools, Machine Bits, Car Bits, Emery, Emery Wheels, Emery Grinders, Vises, Set Screws, Belting, Chucks, Wrenches, Lacing, Lathe Dogs, Bolt Dogs, Iron and Steel Clamps, Babbitt Metal, Machinists'

AND

Railroad Supplies,

OF EVERY DESCRIPTION.

Sole Western Agents for

New York Tap and Die Co.
Tanite Co.'s Emery Wheels, Grinders, &c.
Centerbrook Manuf'g Co. (Bits, &c.)
Henry Whiton, (Chucks.)

Manufacturers' Agents for

Stephens Patent Vise Co.
Morse Twist Drill Co.
Reynolds & Co. (Set Screws.)
Le Count's Lathe Dogs, &c.

EVERY VARIETY OF BROWN & SHARPE'S MACHINISTS' TOOLS.

RHODES & BRADLEY,

DEALERS IN

SCOTCH AND AMERICAN

PIG IRON,

Coal and Iron Ores,

Office, 154 Washington St., near La Salle. Dock, Ill. Central Railroad Slips.

CHICAGO.

Pumps.

Bolen, Crane & Co.	Newark, N. J.	See Page 62
C. Henry Hall & Co.	New York	" 4
Charles B. Hardick	Brooklyn, N. Y.	" 84
Hart, Ball & Hart	Buffalo, N. Y.	" 136
J. B. Waring	New York	" 24
J. D. West & Co.	" "	" 56
J. J. Walworth	Chicago, Ill.	" 124
J. S. Mundy	Newark, N. J.	" 266
Miller & Smith	New York	" 112
Posts & Kalkman	" "	" 266
Speedwell Iron Works	" "	" 122

Punches and Shears.

Isaac H. Shearman	Philadelphia, Pa.	See Page 80
L. W. Pond	Worcester, Mass.	" 158

Punching and Shearing Machines.

Wm. Sellers & Co.	Philadelphia, Pa.	See Page 64
Wm. B. Bement & Son	"	" 168

Putty.

D. R. Hobart & Co.	New York	See Page 126
Leffingwell & Co.	Cleveland, Ohio	" 190

MEASURES.—Continued.

Ale and Beer Measure. (British).

pints.
2 = 1 = quart.
8 = 4 = 1 gallon.
72 = 36 = 9 = 1 firkin.
144 = 72 = 18 = 2 = 1 kilderkin.
288 = 144 = 36 = 4 = 2 = 1 barrel.
432 = 216 = 54 = 6 = 3 = 1½ = 1 hogshead.
576 = 288 = 72 = 8 = 4 = 2 = 1⅓ = 1 puncheon.
864 = 432 = 108 = 12 = 6 = 3 = 2 = 1½ = 1 butt.

Dry Measure.

The Standard Bushel contains 2150.42 cubic inches, or 77.627013 pounds avoirdupois of pure water at maximum density. Its legal dimensions are 18½ inches Diameter inside, 19½ inches outside, and 8 inches deep; and when heaped, the cone must be 6 inches high, making a heaped bushel equal to 1¼ struck ones.

Pints.	Quarts.	Gallons.	Pecks.	Bushels.	Cubic Inches.
2 =	1 =	.250 =	.125 =	.0315 =	67.2
8 =	4 =	1 =	.5 =	.125 =	268.8
16 =	8 =	2 =	1 =	.25 =	537.6
64 =	32 =	8 =	4 =	1 =	2150.42

Liquid Measure.

The standard gallon measures 231 cubic inches, or 8.33888 lbs. avoirdupois of pure water, at about 39.85 degrees Fahr., the barometer at 30 inches.

gills
4 = 1 pint.
8 = 2 = 1 quart.
32 = 8 = 4 = 1 gallon.
1344 = 336 = 168 = 42 = 1 tierce.
2016 = 504 = 252 = 63 = 1½ = 1 hogshead.
2488 = 672 = 336 = 84 = 2 = 1 1-3 = 1 puncheon.
4032 = 1008 = 504 = 126 = 3 = 2 = 1½ = 1 pipe.
8064 = 2016 = 1008 = 252 = 6 = 4 = 3 = 2 = 1 tun.

Cylinders of the following sizes are closely approximating measures.

	Height, inches.	Diameter, inches.		Height, inches.	Diameter, inches.
Gill	3	1¾	Quart	6	3½
½ pint	3⅝	2¼	Gallon	6	7
Pint	3	3½	10 gallons	15	14

A cubic foot contains 7½ gallons.

COPYING PRESSES A SPECIALTY.

T. Shriver & Co.

333 East 56th St., Near 2d Ave., New York,

MANUFACTURERS OF

COPYING PRESSES

OF ALL SIZES.

Presses for Railroad Use Specially, 10x13, 12x16, 15x20, 17x22, 22x24.

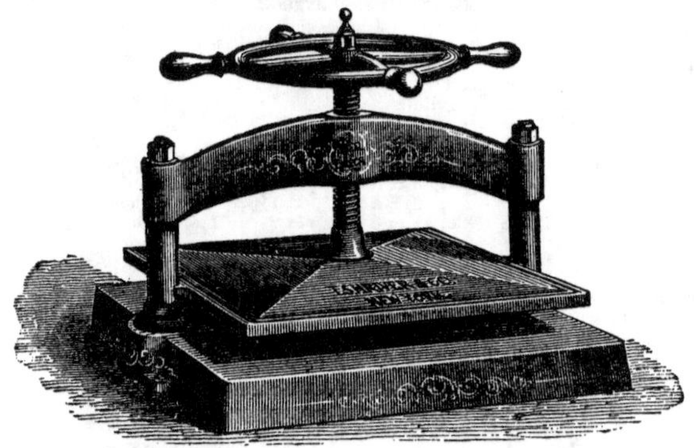

THIS CUT REPRESENTS OUR

LARGE STEEL ARCH COPYING PRESS,

DESIGNED ESPECIALLY FOR

RAILROAD, EXPRESS,

AND

TRANSPORTATION COMPANIES USE.

The *Arch* is of *Steel;* the Bed Plate is braced by our *Patented* Method, giving very great strength; and all the parts are of proportions to bear the heaviest work in copying large WAY BILLS, MANIFESTS, &c.

Address

T. SHRIVER & CO.,

333 East 56th Street, New York.

Quarry Tools.

Nelson Tool Works......New York......See Page 30

Railroad Baggage Checks.

W. W. Wilcox......Chicago, Ill......See Page 190

Railroad Bonds or Securities.

Geo. H. Crain & Co......Chicago, Ill......See Page 124
O. W. Child......New York...... " 266

Railroad Car Forgings.

De Laney & Co......Buffalo, N. Y......See Page 98
Pittsburgh Forge & Iron Co......Pittsburgh, Pa...... " 174

Railroad Castings.

Bowlers, Maher & Brayton......Cleveland, Ohio......See Page 42
Davenport, Fairbairn & Co......Erie, Pa...... " 188
Jonas S. Heartt & Co......Troy, N. Y...... " 134
Ramapo Wheel & Foundry Co......Ramapo, N. Y...... " 256
Wason Manufacturing Co......Springfield, Mass...... " 146

Railroad Chairs.

Pittsburgh Bolt Co......Pittsburgh, Pa......See Page 118

MENSURATION.—2 PAGES.

(*From Hamilton.*)

Mensuration of Surfaces.

Area of any parallelogram...... = base × perpendicular height.
Area of any triangle...... = base × ½ perpendicular height.
Area of any circle...... = diameter ^{2} × .7854.
Area of sector of circle...... = arc × ½ radius.
Area of segment of circle...... = area of sector of equal radius, less area of triangle.
Area of parabola...... = base × 2-3 height.
Area of ellipse...... = longest diameter × shortest diameter × .7854.
Area of cycloid...... = area of generating circle × 3.
Area of any regular polygon...... = sum of its sides × perpendicular from its centre to one of its sides ÷ 2.
Surface of cylinder...... = area of both ends + length × circumference.
Surface of cone...... = area of base + circumference of base × ½ slant height.
Surface of sphere...... = diameter^2 × 3.1415.
Surface of frustrum...... = sum of girt at both ends × ½ slant height + area of both ends.
Surface of cylindrical ring...... = thickness of ring added to the inner diameter × by the thickness × 9.8698.
Surface of segment...... = height of segment × whole circumference of sphere of which it is a part.

POLYGONS.

1. To find the area of any regular polygon. Square one of its sides, and multiply said square by the number in 1st column of the following table.

2. Having a side of a regular polygon, to find the radius of a circumscribing circle. Multiply the side by the corresponding number in the 2d column.

3. Having the radius of a circumscribing circle, to find the side of the inscribed regular polygon. Multiply the radius by the corresponding number in 3d column.

Number of Sides.	Name of Polygon.	1 Area $= S^2 \times$	2 Radius $= S \times$	3 Side $= R \times$	Angle contained betw'n two sides.
3	Equilateral triangle	.433	.5774	1.732	60°
4	Square	1.	.7071	1.4142	90°
5	Pentagon	1.7205	.8507	1.1756	108°
6	Hexagon	2.5891	1.	1.	120°
7	Heptagon	3.6339	1.1524	.8678	128.57°
8	Octagon	4.8284	1.3 66	.7654	135°
9	Nonagon	6.1818	1.4619	.684	140°
10	Decagon	7.6942	1.618	.618	144°
11	Undecagon	9.3656	1.7747	.5635	147.27°
12	Dodecagon	11.1962	1.9319	.5176	150°

In the heads of the columns in above table, S = side, and R = Radius.

MENSURATION.—Continued.

Properties of the Circle.

Diameter × 3.14159 = circumference.
Diameter × .8862 = side of an equal square.
Diameter × .7071 = side of an inscribed square.
Diameter² × .7854 = area of circle.
Radius × 6.28318 = circumference.
Circumference ÷ 3.14159 = diameter.

The circle contains a greater area than any plane figure, bounded by an equal perimeter or outline.

The areas of circles are to each other as the squares of their diameters.

Any circle whose diameter is double that of another contains four times the area of the other.

Area of a circle is equal to the area of a triangle whose base equals the circumference, and perpendicular equals the radius.

Mensuration of Solids.

Cylinder........ = area of one end × length.
Sphere........ = cube of diameter × .5236.
Segment of sphere........ = square root of the height added to three times the square of radius of base × by height and by .5236.
Cone or pyramid........ = area of base × 1-3 perpendicular height.
Frustrum of a cone........ = product of diameter of both ends + sum of their squares, × perpendicular height × .2618.
Frustrum of a pyramid........ = sum of the areas of the two ends + square root of their product, × by 1-3 of the perpendicular height.
Solidity of a wedge........ = area of base × ½ perpendicular height.
Frustrum of a wedge........ = ½ perpendicular height × sum of the areas of the two ends.
Solidity of a ring........ = thickness + inner diameter, × square of the thickness, × 2.4674.

POLYHEDRONS.

No. of sides.	Name.	1 Radius of Circumscribed Circle. R = S×	2 Radius of Inscribed Circle. R = S×	3 Area of Surface. A = S²×	4 Cubic Contents. C = S³×
4	Tetrahedron	.6124	.2041	1.7320	.1178
6	Hexahedron	.866	.5	6.	1.
8	Octahedron	.7071	.4082	3.4641	.4714
12	Dodecahedron	1.4012	1.1135	20.6458	7.6631
20	Icosahedron	.951	.7558	8.6602	2.1817

Side is length of linear edge of any side of the figure.

1.—Radius of Circumscribed Circle = *Side* multiplied by the number in 1st column corresponding to figure.

3.—Radius of Inscribed Circle = *Side* multiplied by the number in 2d column corresponding to figure.

3.—Area of Surface = Square of *side* multiplied by the number in 3d column corresponding to figure.

4.—Cubic Contents = Cube of *side* multiplied by number in 4th column corresponding to figure.

VALENTINE & COMPANY,

MAKERS OF

FINE CAR VARNISHES,

88 CHAMBERS STREET, NEW YORK.

☞ Goods put up in 1 and 5 gallon cans, and barrels of 45 gallons. ☜
In latter shape an allowance is made of 20 cents per gallon, to cover difference in cost of barrels and tins. For cash in 30 days, 5% off.

VALENTINES VARNISHES

FINISHING.

PER GALLON.

Railway Body Varnish $5 50
(For Finishing-coats on Outsides of Cars.)

RUBBING.

Hard Drying Body Varnish 5 00
(For under-coats outside, and finishing inside; also, for finishing Locomotives)

Quick Leveling Varnish 4 00
(For under-coats on inside work and Locomotives; also for quickening the Railway Body and Hard Drying Body.)

MISCELLANEOUS.

No. 1 Black Varnish 1 50
(For smoke-stacks and other iron work.)

Japan Gold-Size 3 50
(For binding, drying and hardening colors.)

Crown Coach Japan 1 75
(For drying and hardening paints.)

Ground Roughstuff 3 00
(For producing a hard and level surface on bodies of Cars.)

Dark Permanent Wood-Filling 4 00
(A patent article for priming outside work, and permanently filling and darkening natural woods.)

Light Permanent Wood-Filling 4 00
(Precisely same as above, but being light colored, it fills without staining the natural wood, and is specially adapted to priming Cars.)

Railroad Iron.

Geo. H. Crain & Co.	Chicago, Ill.	See Page 124
Jno. A. Griswold & Co.	Troy, N. Y.	" 50
Jno. W. Quincy	New York	" 178
Miller & Smith	" "	" 112
O. W. Child	New York	" 262
P. H. Moffat	Chicago, Ill.	" 190
Reese, Graff & Woods	Pittsburgh, Pa.	" 68
Vose, Dinsmore & Co.	New York	" 60
Winans & Co.	" "	" 106

Railroad Machinery.

Schenectady Locomotive Works	Schenectady, N. Y.	See Page 148

Railroad Spike Mauls.

Hermann Boker & Co.	New York	See Pages 152, 154
Skinner & Gifford Manufacturing Co.	" "	See Page 140

Railroad Spikes.

Collier & Scranton; Oxford Iron Co.	New York	See Page 122

See also, Spikes.

FOREIGN MONEY, IN U. S. CURRENCY.

(*Condensed from Trautwine.*)

Crown. Great Britain	$1.13
" Portugal	1.12
" Baden, Bavaria, North Germany	1.06
Copeck. Russia	.0075
Dollar. Chili, Mexico, Peru, Bolivia, New Granada	1.00
" Central America	.95
" Norway, Sweden, Spain	1.05
" Rigsbank Dollar, Denmark	.52
" Specie Dollar, Denmark	1.05
Doubloon. Spain, Mexico	15.65
" New Granada	15.34
Drachm. Greece	.18
Ducat. Austria, Bohemia, Hamburg, Hanover	2.28
" Sweden	2.20
" Denmark	1.81
Franc. France, Belgium	.19
5 Franc piece	.95
Florin. Austria, Silesia	.48
" Holland, Netherlands, South Germany	.40
" (gold) Hanover	1.66
" (silver) "	.56
" Prussia	.55
Guilder. Netherland	.40
Gulden. Baden	.40
Guinea. Great Britain	5.08
Groschen. Prussia	.024
5 Groschen. "	.12
Grote. Bremen	.01
Kreutzer. Bavaria	.0067
10 Kreutzer. Austria	.08
Livre. France, Sardinia	.185
" Tuscany, Venice	.16
Lira. Milan	.14
Mark. Denmark	.09
Milrea. Portugal	1.12
Moidore "	6.50
Mohur. Bombay	7.28
" Bengal	8.15
Ounce. Sicily	2.50
Pistola. Rome	3.37
Pistole. Spain	3.90
Paolo. Rome	.10
Peseta. Spain	.20
Pistareen. Spain	.20
Piastre. Spain	1.04
" Egypt	.05
Para	.009
Pound. Great Britain	4.84
" Canada, New Scotia, New Brunswick, Newfoundland	4.00
Reale plate. Spain	.10
" vellon. "	.05
" South America	.0575
2 Reales. Equador	.1875
Rix dollar. Hamburg, Hanover	1.10
" Sweden, Holland	1.05
" Baden, Brunswick	1.00
" Bavaria, Austria, Hungary	.97
Rix 10 thalers. Prussia	8.00
Rixthaler. Prussia, Poland, N. Germany. Bremen, Saxony, Hanover	.69
Rix species thaler. Saxony	.98
Rupee. Bombay, Madras	.45
" Sicca	.53
Reis (1200), Brazil	.99
Rouble. Russia	.79
Schilling. Hamburg	.02
Shilling. Great Britain	.23
Skilling. Denmark	.0075
Scudo. Piedmont	1.36
" Naples, Sicily	.95
" Sardinia	.92
" Rome	1.00
" Genoa	1.28
Soverain. Austria, Bohemia	3.57
Sovereign. Great Britain	4.86
Sous. France nearly	.01
Stiver. Holland "	.02
Teston. Rome	.30
Testoon. Portugal	.12
Zecchin. Turkey	1.40
Zecchina. Rome	2.27

Chicago Spring Works,

MANUFACTURERS OF

EXTRA TEMPERED

Light Elliptic Cast Steel Springs

FOR CARS AND LOCOMOTIVES.

(PER SPECIFICATION)

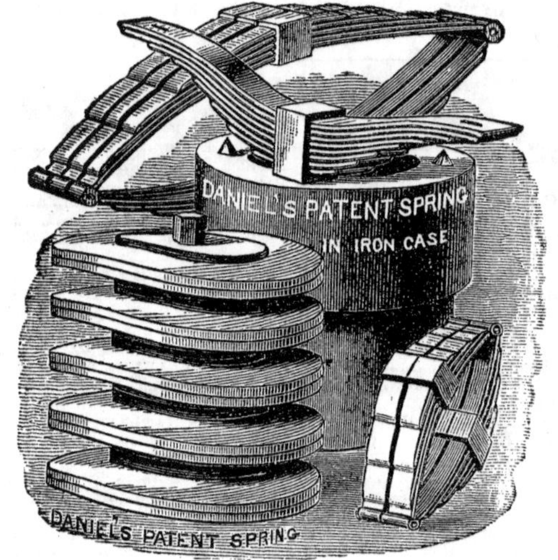

AND THE

Daniel's Patent Car Spring,

FOR FREIGHT CARS.

Rates and Terms as low as can be made. Every Spring *fully tested and warranted,* and orders always promptly filled.

F. M. ATKINSON, Pres't and Sup't.

OFFICE and WORKS:

No. 235 South Clinton Street, Chicago.

Railroad Supplies.

Railroad Tools.

Railroad Tickets.

STEEL SPRINGS.

(Hamilton.)

Rule 1st—To find elasticity of a given steel-plate spring; Breadth of plate in inches multiplied by cube of the thickness in $\frac{1}{16}$ inch, and by number of plates; divide cube of span in inches by prod ct so found, and multiply by 1.66. Result, equal elasticity in $\frac{1}{16}$ of an inch per ton of load.

Rule 2d—To find span due to a given elasticity, and number and size of plate; Multiply elasticity in sixteenths per ton, by breadth of plate in inches, and divide by cube of the thickness in inches, and by the number of plates; divide by 1.66 and find cube root of the quotient. Result, equal span in inches.

Rule 3d—To find number of plates due to a given elasticity, span and size of plates: Multiply the cube of the span in inches by 1.66; multiply the elasticity in sixteenths by the breadth of the plate in inches, and by the cube of the thickness in sixteenths; divide the former product by the latter. The quotient is the number of plates.

Rule 4th—To find working strength of a given steel-plate spring: Multiply the breadth of plate in inches by the square of the thickness in sixteenths, and by the number of plates; multiply also the working span in inches by 11.3; divide the former product by the latter. Result, equal working strength in tons burden.

Rule 5th—To find span due to a given strength and number, and size of plate; Multiply the breadth of plate in inches by the square of the thickness in sixteenths, and by the number of plates; multiply, also, the strength in tons by 11.3; divide the former product by the latter. Result equal working span in inches.

Rule 6th—To find the number of plates due to a given strength, span and size of plate: Multiply the strength in tons by span in inches, and divide by 11.3; multiply also the breadth of plate in inches by the square of the thickness in sixteenths; divide the former product by the latter. Result, equal number of plates.

The span is that due to the form of the spring loaded. Extra thick plates must be replaced by an equivalent number of plates of the ruling thickness, before applying the rule. To find this, multiply the number of extra plates by the square of their thickness, and divide by the square of the ruling thickness; conversely, the number of plates of the ruling thickness to be removed for a given number of extra plates, may be found in the same way.

Railroad Track Scales.

Buffalo Scale Company........................Buffalo, N. Y..................See Page 40
Jones Scale Works........................Binghamton, N. Y.......... " 92

Rails.

Chas. W. Matthews........................Philadelphia, Pa...........See Page 38
Greene & Randolph........................New York.................... " 54
Jones & Laughlins........................Pittsburgh, Pa................ " 32
Jno. W. Quincy........................New York.................... " 178
Miller & Smith........................ " " .... " 112
O. W. Child........................ " " .... " 262
Tyng & Co........................ " " .... " 260
Vose, Dinsmore & Co........................ " " .... " 60

Railway Brakes.

Ward Air Brake Co........................Kalamazoo, Mich...........See Page 56

Ratchets.

Barwick Wrench Co........................Boston, Mass.................See Page 60

Reamers.

Champlin & Rogers........................Chicago, Ill...................See Page 192

Red Lead.

Atlantic White Lead Co.—Rob't Colgate & Co. New York...........See Pages 238, 240
Brooklyn White Lead Co........................ " " ....See Page 268
T. B. Bickerton & Co........................Philadelphia, Pa............. " 136
Wetherill & Bro........................Philadelphia, Pa............. " 52

Reflectors.

Radley & McAlister Manufacturing Co...... .....New York.....................See Page 250
Vose, Dinsmore & Co........................ " " .... " 60

Retorts.

Philip Neukumet........................Philadelphia, Pa...........See Page 104
Maurer & Weber........................New York.................... " 228

Riddles.

Van Tuyl Manufacturing Co........................New York.....................See Page 242

Rivet Hammers.

Hermann Boker & Co........................New York.............See Pages 152, 154
Nelson Tool Works........................ " " ....See Page 30

Rivet Makers.

Hoopes & Townsend........................Philadelphia, Pa...........See Page 28
J. T. Ryerson........................Chicago, Ill........... .... " 180
Pittsburgh Bolt Co........................Pittsburgh, Pa............... " 118
Providence Tool Co.; H. B. Newhall, Agent....New York.................... " 165

ROOFING SLATE.

General Rule for the Computation of Slate.

(Furnished by the Penrhyn Slate Co.)

From the length of the Slate take three inches, or as many as the third covers the first; divide the remainder by 2, and multiply the quotient by the width of the slate, and the product will be the number of square inches in a single slate. Divide the number of square inch s thus procured by 144, the number of square inches in square foot, and the quotient will be the number of feet and inc es required. A square of slate is what will cover 100 feet square, when properly laid upon the roof.

TABLE OF SIZES AND NUMBER OF SLATE IN ONE SQUARE.

Size in Inches.	No. of Slate in a Square.	Size in Inches.	No. of Slate in a Square.	Size in Inches.	No. of Slate in a Square.	Size in Inches.	No. of Slate in a Square.
6×12 ...	533	9×14 ...	291	10×18 ...	192	11×22 ...	137
7×12 ...	457	10×14 ...	261	11×18 ...	174	12×22 ...	126
8×12 ...	400	12×14 ...	218	12×18 ...	160	14×22 ...	108
9×12 ...	355	8×16 ...	277	14×18 ...	137	12×24 ...	114
10×12 ...	320	9×16 ...	246	10×20 ...	169	14×24 ...	98
12×12 ...	266	10×16 ...	221	11×20 ...	154	16×24 ...	86
7×14 ...	374	12×16 ...	184	12×20 ...	141	14×26 ...	89
8×14 ...	327	9×18 ...	213	14×20 ...	121	16×26 ...	78

THE WEIGHT of a Square of Slate is estimated in a general way (varying according to the thickness of the different makes), at from 600 to 700 lbs. per square.

Rivets.

Bickford, Curtiss & Deming.......Buffalo, N. Y.......See Page 136
D. Brewer & Co.......Philadelphia, Pa.......“ 246
N. H. Gardner & Co.......Buffalo, N. Y.......“ 254
Wm. Green & Co.......Wilmington, Del.......“ 248
See also, Rivet Makers.

Riveting Machines.

Wm. Sellers & Co.......Philadelphia, Pa.......See Page 64

Rock Drills.

Rand & Waring Drill & Compressor Co.......New York.......See Front Page

Rods—Measuring.

F. Eckel.......New York.......See Page 76

Rolling Mill Castings.

Bowlers, Maher & Brayton.......Cleveland, Ohio.......See Page 42

Rolling Mill Machinery.

Phelps & Sanger.......Cleveland, Ohio.......See Page 120

Roofing Materials.

H. W. Johns.......New York.......See Page 224

Roofing Plates.

Tyng & Co.......New York.......See Page 260

LOCKS AND PADLOCKS.

STANDARD SIZES.

(*Furnished by Romer & Co.*)

BRASS SPRING PADLOCKS.

No.
0, 1 inch Spring Padlock, no Drop.
½, 1⅛ “ “ “
1, 1¼ “ “ “
2, 1½ “ “ “
3, 1¾ “ “ “
4, 2⅛ “ “ “
5, 2½ “ “ “
6, 2⅛ “ “ plain drop.
7, 2½ “ “ “
8, 2⅛ “ “ spring drop
9, 2½ “ “ “
10, 3 “ “ “
11, 2½ “ Spring Padlock, spring drop and chain.
12, 2⅛ “ Spring Padlock, spring drop and chain.
13, 2½ “ Spring Padlock, p'ain drop and chain.
14, 2⅛ “ Spring Padlock, plain drop and chain.
15, 2½ “ heavy R. R. Freight Car Lock, spring drop and chain.
16, 2½ inch heavy Spring Padlock, spring drop, no chain.
17, 2⅝ inch Spring Padlock, spring drop, no chain.
18, 2⅝ inch heavy R. R. Car Lock, spring drop and chain.
19, 3 inch heavy R R. Car Lock, spring drop and chain.
20, 3 inch heavy, 3 tumbler R. R. Car Lock and chain.
21, 3 inch heavy 3 tumbler Store Door Padlock, spring drop, no chain.
22, 2½ inch R. R. Switch and Car Lock, spring drop and chain.
23, 2¼ inch 3 tumbler, Spring Lock, spring drop, no chain.
24, 2½ inch 3 tumbler, Spring Lock, spring drop, no chain.
25, 2⅝ inch R R. Freight Car Tumbler Lock, spring drop and chain.

IRON SPRING PADLOCKS.

No.
26, 2¼ inch Spring Padlock, no drop.
27, 2¼ “ “ brass drop.
28, 2⅝ “ “ no drop.
29, 2⅝ “ “ brass drop.
30, 2⅝ “ Spring Padlock, brass drop and chain.

SELF-LOCKING TUMBLER BRASS SPRING PADLOCK.

(*Drilled Keys.*)

No.
60, 2 in. Lock, spring drop, no chain.
61, 2 “ “ and “
62, 2¼ “ “ no “
63, 2¼ “ “ and “
64, 2½ “ “ no “
65, 2½ “ “ and “
66, 2¾ “ “ no “
67, 2¾ “ “ and “
Extra Keys, $1.50 per dozen.

SELF-LOCKING TUMBLER BRASS SPRING PADLOCK.

(*Flat Keys.*)

No.
68, 2 inch Lock, no chain.
69, 2 “ “ with “
70, 2¼ “ “ no “
71, 2¼ “ “ with “
72, 2½ “ “ no “
73, 2½ “ “ with “
74, 2¾ “ “ no “
75, 2¾ “ “ with “

IMPROVED PASSENGER CAR DOOR LOCKS.

Iron Case, brass inside works and key.
Brass Case, “ “ “
Silver Plated Case, brass inside works aud key.
Silver Plated Case, with plated knobs, brass inside works and key.

A. L. HALE AND BRO.

Wholesale and Retail

Furniture Dealers,

10, 12, 14 AND 16 CANAL STREET,

CHICAGO, ILLINOIS.

☞ We have the largest and best stock of Furniture in the West, and make a specialty of

OFFICE FURNISHING.

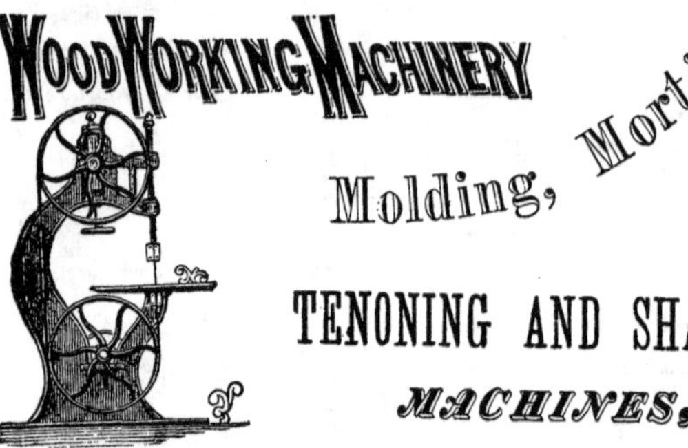

WOOD WORKING MACHINERY

Mortising,

Molding,

TENONING AND SHAPING

MACHINES,

BAND SAWS, SCROLL SAWS,

Planing and Matching Machines, &c.,

FOR RAILROAD, CAR AND AGRICULTURAL SHOPS.

SUPERIOR TO ANY IN USE.

J. A. FAY & CO., CINCINNATI, OHIO.

Roofing Slates.

Penrhyn Slate Co.........Buffalo, N. Y.........See Page 202

Ropes.

De Grauw, Aymar & Co.........New York.........See Page 112
T. B. Bickerton & Co.........Philadelphia, Pa......... " 138

Rubber Belting, Hose, &c.

A. Carr.........New York.........See Page 244
A. Fulton's Son & Co.........Pittsburgh, Pa......... " 116
Crane Bros. Manufacturing Co.........Chicago, Ill......... " 230
Goodyear's I. R. G. Manufacturing Co.........New York......... " 252
J. J. Walworth.........Chicago, Ill......... " 124
New York Rubber Co.........New York......... " 260
Phelps & Sanger.........Cleveland, Ohio......... " 120
Radley & McAlister Manufacturiug Co.........New York......... " 250
Vose, Dinsmore & Co......... " "......... " 60

Rubber Goods.

Goodyear's I. R. G. Manufacturing Co.........New York.........See Page 252
L. G. Tillotson & Co......... " "......... " 94
New York Rubber Co......... " "......... " 260
Phelps & Sanger.........Cleveland, Ohio......... " 120

Rubber Springs.

Chicago Spring Works.........Chicago, Ill.........See Page 200
Goodyear's I. R. G. Manufacturing Co.........New York......... " 252
Greene & Randolph......... " "......... " 54
L. G. Tillotson & Co......... " "......... " 94
New York Rubber Co......... " "......... " 260
Radley & McAlister Manufacturing Co......... " "......... " 250
Vose, Dinsmore & Co......... " "......... " 60

Safes.

W. A. Dobinson & Co.........Buffalo, N. Y.........See Page 196
Warren, Johnson & Co......... "......... " 100

Safety Fuse.

F. M. Alford.........Avon, Conn.........See Page 158
Geo. W. Mowbray.........North Adams, Mass......... " 182
Laflin & Rand Powder Co.........New York.........See Last Page

Safety Valves.

Geo. W. Richardson & Co.........Troy, N. Y.........See Page 50
Posts & Kalkman.........New York......... " 266

Sandpaper, &c.

England & Bindley.........Pittsburgh, Pa.........See Page 162

Sash and Blind Machinery.

C. B. Rogers & Co.........New York.........See Page 210

Sashes, Doors, &c.

Joseph Churchyard.........Buffalo, N. Y.........See Page 160

WOOD-WORKING MACHINERY.

Velocity of circular-saw at periphery, 6,000 to 7,000 feet per minute.
" the band-saw, 2,500 feet per minute.
" gang-saws, 20-inch stroke, 120 strokes a minute.
" scroll-saws, 300 strokes per minute.
" planing-machine cutters at periphery, 4,000 to 6,000 feet per minute.
Travel of work under planing-machine, 1-20th of an inch for each cut.
" planing-machine cutters, 3,500 to 4,000 feet per minute.
" squaring-up-machine cutters, 7,000 to 8,000 feet per minute.
Speed of wood-carving drills, 5,000 revolutions per minute.
" machine augers, 1½ inch diameter, 900 revolutions per minute.
" machine augers, ¾ inch diameter, 1,200 revolutions per minute.
Gang-saws require for 45 super. feet of pine per hour, 1 HP indicated.
Circular-saws require for 75 super. feet of pine per hour, 1 HP indicated.
In oak or hard wood, ¾ths of the above quantity require 1 HP indicated.

LEATHER GLUE.

A substance known as "leather glue" is prepared by mixing ten parts of sulphide of carbon with one of oil of turpentine, and adding enough gutta-percha to thicken the mass. The leather surfaces to be united must be freed from oil, which is accomplished by subjecting them to pressure by laying the leather upon blotting paper and applying a hot iron. After tracking together the edges to be joined with the cement, they are to be kept under pressure until the glue is entirely dry.

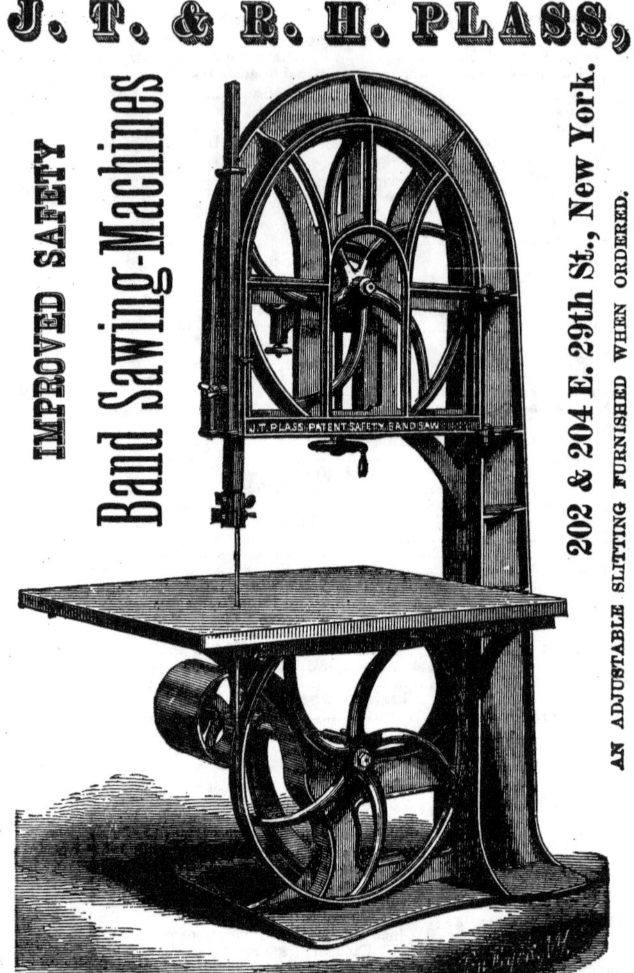

In all other Band Sawing-Machines the saw is placed outside of the frame and column which support the driving parts, exposing the Saws to injury from handling lumber about them, also rendering them extremely dangerous to the operator. The main feature in this Machine has been to produce a Band Saw, perfectly safe to the operator, and fully protected from injury to the Saw, combining neatness and utility in all its parts. The Saw is placed to run upward near the column, under the top arch, over the upper or tension pulley, and down near the front frame and behind the guide-bar, which can be adjusted up and down to the required height for the work to be done. It also runs through a superior adjustable guide, above and below the work, and a guide on the column, guiding equally well all widths of Saws.

Sash Weights.

Crane Bros. Manufacturing Co. .... Chicago, Il. .... See Page 230

Saw Gummers.

D. Brewer & Co. .... Philadelphia, Pa. .... See Page 246

Saw Frames.

E. M. Boynton .... New York .... See Page 6

Saw Handles.

E. M. Boynton .... New York .... See Page 6

Saw Makers.

American Saw Company .... New York .... See Page 156
C. Van Ness .... " " .... " 70
E. M. Boynton .... " " .... " 6

See also, Saws.

Saw Mill Machinery.

Snyder Bros .... Williamsport, Pa. .... See Page 46

Saw Mills.

Buffalo Machinery Agency .... Buffalo, N. Y. .... See Page 160

Saws.

American Saw Co .... New York .... See Page 156
C. Van Ness .... " " .... " 70
E. M. Boynton .... " " .... " 6
J. T. & R. H. Plass .... " " .... " 208
Pratt & Co .... Buffalo, N. Y. .... " 132

Scales.

A. M. Gilbert & Co .... Chicago, Ill .... See Page 181
Buffalo Scale Co .... Buffalo, N. Y. .... " 40
Jones Scales Works .... Binghamton, N. Y. .... " 92
Phelps & Sanger .... Cleveland, Ohio .... " 120
W. A. Dobinson & Co .... Buffalo, N. Y. .... " 196
Wm. P. Kellogg & Co .... Troy, N. Y. .... " 98

Scientific Books.

D. Van Nostrand .... New York .... See Page 258

Scrapers.

Skinner & Gifford Manufacturing Co .... Dunkirk, N. Y. .... See Page 140

Scrap Iron.

Atwater, Wheeler & Co .... New Haven, Conn .... See Page 10
Chas. W. Matthews .... Philadelphia, Pa. .... " 38

Screw Bolts.

Lock Nut & Bolt Co. of New York .... New York .... See Page 272
Providence Tool Co; H. B. Newhall, Agent ..... " " .... " 166

TO JOIN THE END OF BAND SAWS.

Furnished by J. T. & R. H. Plass, New York.

File the ends of the band on opposite sides, to form two wedged-shaped ends, having a lap of say ¾ inch long, which, when laid with their beveled and filed sides together shall form a good joint of the same thickness as blade. Now clamp the ends on a piece of board, with the back of the blade toward you, with the lap brought fair together, and see that the back of the blade is straight. Cut a piece of "silver solder," large enough to cover the lap, lay it between the lapped portion with a little pulverized "borax." Now, having a piece cut out of your board, say three inches wide, directly under your lap, heat your soldering tongs to a bright "cherry red," and hold them pinched firmly on the lap until the solder flows freely from under the joint, then cool off the tongs and soldered portion of saw by pouring water upon the same, without relieving the pressure, until nearly cold. Try a file on both sides of the blade, and should it be harder than the other part of the blade, reheat your tongs a little, and draw the temper by pressing the tongs upon the hard portion of blade till partially heated, but not upon the lap, as it will weaken your joint. File off the solder and joint to the same thickness as other parts, and the soldering of your band is completed.

CEMENT FOR LEATHER BELTING.

Take of common glue and American isinglass, equal parts: place them in a boiler and add water sufficient to cover the whole. Let it soak 10 hours, then bring it to a boiling heat, and add pure tannin until the whole becomes ropey or appears like the white of eggs. Apply it warm. Buff the grain off the leather where it is to be cemented; rub the joint surfaces solidly together, let it dry a few hours, and it is ready for practical use; and, if properly put together, it will not need riveting, as the cement is nearly of the same nature as the leather itself.

RULES FOR CALCULATING THE SPEED OF DRUMS AND PULLEYS.

(*Furnished by Jas. S. Nason.*)

PROBLEM No. I.—The diameter of the driven being given, to find its number of revolutions.

Rule—Multiply the diameter of the driver by number of its revolutions, and divide the product by the diameter of the driven, the quotient will be the number of revolutions of the driven.

PROBLEM No. II.—The diameter and revolutions of the driver being given, to find the diameter of the driven that shall make any given number of revolutions in the same time.

Rule—Multiply the diameter of the driver by its number of revolutions, and divide the product by the number of revolutions of the driven; the quotient will be its diameter.

PROBLEM No. III.—To ascertain the size of the driver.

Rule—Multiply the diameter of the driven by the number of revolutions you wish it to make, and divide the product by the revolutions of the driver; the quotient will be the diameter of the driver.

WEIGHT OF GRINDSTONES.

RULE: Square the diameter (in inches), multiply by thickness (in inches), then multiply by decimal .06363.

Example: Find the weight of a stone 4 feet 6 inches diameter and 7 inches thick.

4 ft. 6 in. = 54 inches; square of 54=2916; multiplied by 7=20412; multiplied by .06363=*Ans.*, 1298.815 lbs., which is weight of stone.

Shears and Punches.

L. W. Pond........Worcester, Mass........See Page 158
See also, Punches.

Shearing Machines.

Wm. Sellers & Co........Philadelphia, Pa........See Page 64

Sheet Iron.

Atwater, Wheeler & Co........New Haven, Conn........See Page 10
Bruce & Cook........New York........ " 112
John Merry & Co........ " " .... " 48
J. T. Ryerson........Chicago, Ill........ " 180
Marshall Lefferts, Jr........New York........ " 44
P. H. Moffat........Chicago, Ill........ " 190
Radley & McAlister Manufacturing Co........New York........ " 250
Sidney Shepard & Co........Buffalo, N. Y........ " 196
Wm. Green & Co........Wilmington, Del........ " 248

Sheet Lead.

Bruce & Cook........New York........See Page 112
E. W. Blatchford & Co........Chicago, Ill........ " 124

Sheet Rubber.

A. Fulton's Son & Co........Pittsburgh, Pa........See Page 116
Goodyear's I. R. G. Manufacturing Co........New York........ " 252
New York Rubber Co........ " " .... " 260

Sheet Zinc.

Bruce & Cook........New York........See Page 112

Shingles.

Joseph Churchyard........Buffalo, N. Y........See Page 160

Shingle Strips.

Reese, Graff & Woods........Pittsburgh, Pa........See Page 68

Ship Lanterns.

S. M. Aikman & Co........New York........See Page 114

IRON TELEGRAPH WIRE.

TABLE OF LENGTH, SIZE, WEIGHT, STRENGTH, ETC.

(Charles T. Chester.)

Birmingham Wire Gauge.	Diameter in Inches.	Weight 100 ft.	Weight 1 Mile, Galvanized.	Weight 1 Mile, not Galvan.	Breaking Strain.	Length of Bundles in Ft.
0	0.340	29.44	1490	1416	7280	213
1	0.300	22.92	1210	1150	5650	273
2	0.280	19.97	1054	1002	4930	315
3	0.260	17.22	909	854	4250	363
4	0.240	11.00	775	747	3620	429
5	0.220	12.34	651	619	3040	510
6	0.200	10.19	538	512	2510	609
7	0.185	8.72	461	438	2220	717
8*	0.170	7.37	389	370	1840	858
9*	0.155	6.12	323	307	1560	1026
10*	0.140	4.99	264	251	1280	1260
11*	0.125	3.98	211	200	1000	1587
12	0.110	3.08	163	157	800	2100
13	0.095	2.35	124	118	568	2679
14	0.085	1.84	97	93	456	3426
15	0.075	1.43	76	73	352	4404
16	0.065	1.08	57	55	264	5862
17	0.057	0.83	44	42	208	7620
18	0.050	0.64	34	32	160	9450
19	0.045	0.52	27	26	128	12255
20	0.040	0.41	21	20	104	14736

* Those marked with star are standard sizes for telegraph use.

Shovels.

England & Bindley....Pittsburgh, Pa....See Page 162
Geo. Worthington & Co....Cleveland, Ohio.... " 100
L. G. Tillotson & Co....New York.... " 94
T. B. Bickerton & Co....Philadelphia, Pa.... " 138

Signal Oil.

Eclipse Lubric Oil Co....New York....See Page Facing Title.
F. S. Pease....Buffalo, N. Y....See Page 100
P. H. Moffat....Chicago, Ill.... " 190
Wm. B. Browne & Co....New York.... " 266

Signals.

Buffalo Steam Gauge Co....Buffalo, N. Y....See Page 170

Slate.

Penrhyn Slate Co....Buffalo, N. Y....See Page 202

Slate Roofing Tools.

Nelson Tool Works....New York....See Page 30

Sledges.

Nelson Tool Works....New York....See Page 30
Wm. Green & Co....Wilmington, Del.... " 248

Slide Lathes—Self-Acting.

Wm. Sellers & Co....Philadelphia, Pa....See Page 64

Sliding Tables.

Wm. Sellers & Co....Philadelphia, Pa....See Page 64

Slotting Machines.

Isaac H. Shearman....Philadelphia, Pa....See Page 80
Wm. Sellers & Co.... " " .... " 64
Wm. B. Bement & Son.... " " .... " 168

Smelters.

Du Plaine & Reeves....Philadelphia, Pa....See Page 102
Hooks Smelting Co.... " " .... " 216
Philadelphia Smelting Co.... " " .... " 134

WROUGHT IRON WELDED TUBES, EXTRA STRONG.

Table of Standard Sizes.—(Morris, Tasker & Co.)

Nominal Diameter.	Actual Outside Diameter.	Thickness, Extra Strong.	Thickness, Double Extra Strong.	Actual Inside Diameter, Extra Strong.	Actual Inside Diameter, Double Extra Strong.
1/8	.405	.100	—	.205	—
1/4	.54	.123	—	.294	—
3/8	.675	.127	—	.421	—
1/2	.84	.149	.298	.542	.244
3/4	1.05	.157	.314	.736	.422
1	1.315	.182	.364	.951	.587
1¼	1.66	.194	.388	1.272	.885
1½	1.9	.203	.406	1.494	1.088
2	2.375	.221	.442	1.933	1.491
2½	2.875	.280	.560	2.315	1.755
3	3.5	.304	.608	2.892	2.284
3½	4.	.321	.642	3.358	2.716
4	4.5	.341	.682	3.818	3.136

THERMOMETERS.

To convert degrees, Centigrade or Reaumur, into degrees Fahrenheit.

Let F = No. of degrees Fahrenheit.
C = " " Centigrade.
R = " " Reaumur.

$$F = \frac{9C}{5} + 32 \qquad F = \frac{9R}{4} + 32 \qquad F = C + R + 32$$

$$C = \frac{4(F-32)}{9}. \qquad R = \frac{4(F-32)}{9}.$$

Freezing point, or 32° Fah. = Zero in Cent., or Reaumur.
Boiling point, or 212° Fah. = 100° Cent., or 80° Reaumur.

W. KUEBLER. F. SEELHORST.

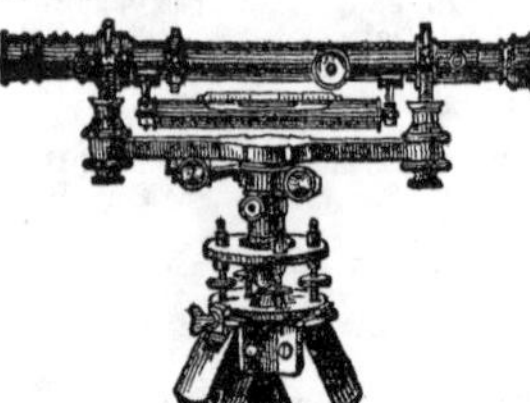

KUEBLER & SEELHORST,

MANUFACTURERS OF

Mathematical

AND Optical

INSTRUMENTS,

No 27 South Eighth Street, cor. Jayne,

(Entrance on Jayne Street.) PHILADELPHIA.

We call the attention of Engineers to our Patented Improvement on Telescopes.

HOOKS SMELTING CO.,

SUCCESSORS TO

H. W. HOOK.
METALLURGIST.
SMELTER & BRASS FOUNDER
MANUFACTURER OF
TYPE. STEREOTYPE.
BABBIT & ANTI-FRICTION
METALS.
ALSO IMPORTER OF
BLOCK TIN. LEAD.
ANTIMONY &c.
BROAD & HAMILTON ST. PHILA.
& 54 WILLIAM ST. NEW YORK.

BLACK DIAMOND FILE WORKS

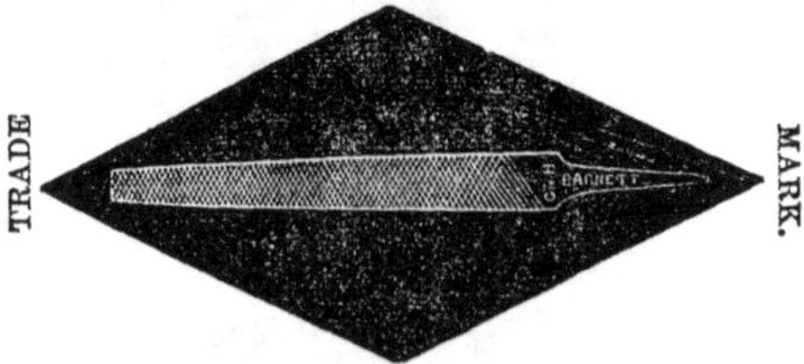

G. & H. BARNETT,

39, 41 & 43 RICHMOND ST., PHILADELPHIA.

Smiths' Tools.

England & Bindley	Pittsburgh, Pa	See Page 162
Nelson Tool Works	New York	" 30
S. S. Townsend	New York	" 2

Smoke Stacks.

Radley & McAlister Manufacturing Co	New York	See Page 250

Socket Pipes.

R. A. Brick & Co	New York	See Page 48

Solder.

Du Plaine & Reeves	Philadelphia, Pa	See Page 102
Holmes & Lissberger	New York	" 84
Lucius Hart & Co	" "	" 56

Spades.

Vose, Dinsmore & Co	New York	See Page 60

Spark Arresters.

Radley & McAlister Manufacturing Co	New York	See Page 250

Spelter.

Holmes & Lissberger	New York	See Page 84
Lucius Hart & Co	" "	" 56
Philadelphia Smelting Co	Philadelphia, Pa	" 134
Wm. Green & Co	Wilmington, Del	" 248

Sperm Oil.

F. S. Pease	Buffalo, N. Y	See Page 100
P. H. Moffat	Chicago, Ill	" 190
Wm. B. Browne & Co	New York	" 266

TELEGRAPH MAGNET AND OFFICE WIRES.

TABLE SHOWING THE NUMBER OF FEET BARE COPPER WIRE TO THE POUND.

(*Chas. T. Chester.*)

Birmingham Wire Gauge.	Diameter in Ins.	No. of Feet Bare Wire in lb.	Birmingham Wire Gauge.	Diameter in Ins.	No. of Feet Bare Wire in lb.
1	.300	3.82	21	.035	280.1
2	.280	4.41	*22	.030	381.6
3	.260	5.09	*23	.027	450.0
4	.240	6.00	*24	.025	533.1
5	.220	7.11	25	.023	630.0
6	.200	8.61	*26	.019	923.4
7	.185	10.09	27	.018	1029.0
8	.170	11.91	*28	.016	1302.0
9	.155	14.32	29	.015	1481.4
10	.140	17.50	*30	.014	1708.5
11	.125	22.1	31	.012	2314.8
*12	.110	28.4	*32	.010	3333.3
*13	.095	37.2	33	.009	3421.4
*14	.085	48.0	*34	.0096	3616.8
15	.075	61.3	35	.0087	4399.8
*16	.065	81.9	36	.0079	5349.0
17	.057	106.1	37	.0067	7425.6
*18	.050	137.9	38	.0058	9908.7
19	.045	161.3	39	.0042	18896.1
*20	.040	215.6	40	.0039	21915.0

* Standard Sizes.

The following are the conducting powers of the several Coppers, 100 being the standard for chemically pure:

Lake Superior Native, not fused	98.8	at 15.5
" " fused, as it comes in commerce	92.6	" 15.0
Burra Burra	88.7	" 14.0
Best Selected	81.3	" 14.2
Bright Copper Wire	72.2	" 15.7
Tough Copper	71.0	" 17.3
Demidoff	59.3	" 12.7
Rio Tinto	14.2	" 14.8

Spiegeleisen.

Hermann Boker & Co....New York....See Page 152, 154

Spike Mauls.

Skinner & Gifford Manufacturing Co....Dunkirk, N. Y....See Page 140

Spikes.

Collier & Scranton; Oxford Iron Co.	New York	See Page 122
England & Bindley	Pittsburgh, Pa.	" 162
G. B. Walbridge	New York	" 130
Geo. Worthington & Co	Cleveland, Ohio	" 100
Greene & Randolph	New York	" 54
Jno. W. Quincy	" "	" 178
Jones & Laughlins	Pittsburgh, Pa.	" 32
Lock Nut & Bolt Co. of New York	New York	" 272
Marshall Lefferts, Jr.—(Galvanized Spikes)	" "	" 44
Miller & Smith	" "	" 112
O. W. Child	" "	" 262
Pratt & Co—(Rail, Bridge, and Boat)	Buffalo, N. Y.	" 132
Wm. Green & Co	Wilmington, Del.	" 248

SIDES OF SQUARES.

Equal in Area to a Circle of any Diameter.

(Condensed from Haswell.)

Diam. of Circle.	Side of Equal Square.	Diam. of Circle.	Side of Equal Square.	Diam. of Circle.	Side of Equal Square.	Diam. of Circle.	Side of Equal Square.	Diam. of Circle.	Side of Equal Square.
1	.886	21	18.611	41	36.335	61	54.06	81	71.784
½	1.329	½	19.054	½	36.778	½	54.503	½	72.227
2	1.772	22	19.497	42	37.221	62	54.946	82	72.67
½	2.215	½	19.94	½	37.664	½	55.389	½	73.113
3	2.658	23	20.383	43	38.108	63	55.832	83	73.557
½	3.102	½	20.826	½	38.551	½	56.275	½	74.
4	3.545	24	21.269	44	38.994	64	56.718	84	74.443
½	3.988	½	21.712	½	39.437	½	57.161	½	74.886
5	4.431	25	22.155	45	39.88	65	57.604	85	75.329
½	4.874	½	22.599	½	40.323	½	58.048	½	75.772
6	5.317	26	23.042	46	40.766	66	58.491	86	76.215
½	5.76	½	23.485	½	41.209	½	58.934	½	76.658
7	6.203	27	23.928	47	41.652	67	59.377	87	77.101
½	6.646	½	24.371	½	42 096	½	59.820	½	77.545
8	7.09	28	24.814	48	42.584	68	60.263	88	77.988
½	7.533	½	25.257	½	42.982	½	60.706	½	78.431
9	7.976	29	25 7	49	43.425	69	61.149	89	78.874
½	8.419	½	26.143	½	43.868	½	61.593	½	79.317
10	8.862	30	26.587	50	44.311	70	62.036	90	79.76
½	9.305	½	27.03	½	44.754	½	62.479	½	80.203
11	9.748	31	27.473	51	45.197	71	62.922	91	80.646
½	10.191	½	27.916	½	45.64	½	63.365	½	81.09
12	10.634	32	28.3 9	52	46.084	72	63.808	92	81.533
½	11.078	½	28.802	½	46.527	½	64.251	½	81.976
13	11.521	33	29.245	53	46.97	73	64.694	93	82.419
½	11.964	½	29.688	½	47.413	½	65.137	½	82.862
14	12.407	34	30.131	54	47.856	74	65.581	94	83.305
½	12.85	½	30.575	½	48.299	½	66.024	½	83.748
15	13.293	35	31.018	55	48.742	75	66.467	95	84.191
½	13.736	½	31.461	½	49.185	½	66.91	½	84.634
16	14.179	36	31 904	56	49.628	76	67.353	96	85.078
½	14.622	½	32.347	½	50.072	½	67.796	½	85.521
17	15.066	37	32.790	57	50.515	77	68.239	97	85.964
½	15.509	½	33.233	½	50.958	½	68.682	½	86.407
18	15.952	38	33.676	58	51.401	78	69.125	98	86.85
½	16.395	½	34.119	½	51.844	½	69.569	½	87.293
19	16.838	39	34.563	59	52.287	79	70.012	99	87.736
½	17 281	½	35.006	½	52.73	½	70.455	½	88.179
20	17.724	40	35.449	60	53.173	80	70.898	100	88.622
½	18.167	½	35.892	½	53.616	½	71.341	½	89.066

CIRCUMFERENCES AND AREAS OF CIRCLES.

Diam.	Circ.	Area.	Diam.	Circ.	Area.	Diam.	Circ.	Area.
1-32—	.0981	.00076	3¼ –	10.21	8.295	7½—	23.56	44.178
1-16	.1963	.00306		10.60	8.946		23.95	45.663
1-8 –	.3926	.01227	3½—	10.99	9.621	7¾ –	24.34	47.173
3-16	.5890	.02761		11.38	10.320		24.74	48.707
1-4—	.7854	.04908	3¾ –	11.78	11.044	8 —	25.13	50.265
5-16	.9817	.07669		12.17	11.793		25.52	51.848
3-8 –	1.178	.1104	4 —	12.56	12.566	8¼ –	25.91	53.456
7-16	1.374	.1503		12.95	13.364		26.31	55.088
1-2—	1.570	.1963	4¼ –	13.35	14.186	8½—	26.70	56.745
9-16	1.767	.2485		13.74	15.033		27.09	58.426
5-8 –	1.963	.3067	4½—	14.13	15.904	8¾ –	27.48	60.132
11-16	2.159	.3712		14.52	16.800		27.88	61.862
3-4—	2.356	.4417	4¾ –	14.92	17.720	9 —	28.27	63.617
13-16	2.552	.5184		15.31	18.665		28.66	65.396
7-8 –	2.748	.6013	5 —	15.70	19.635	9¼ –	29.05	67.200
15-16	2.945	.6902		16.10	20.629		29.45	69.029
1 —	3.141	.7854	5¼ –	16.49	21.647	9½—	29.84	70.882
	3.534	.9940		16.88	22.690		30.23	72.759
1¼ –	3.927	1.227	5½—	17.27	23.758	9¾ –	30.63	74.662
	4.319	1.484		17.67	24.850		31.02	76.588
1½—	4.712	1.767	5¾ –	18.06	25.967	10 —	31.41	78.539
	5.105	2.073		18.45	27.108		31.80	80.515
1¾ –	5.497	2.405	6 —	18.84	28.274	10¼ –	32.20	82.516
	5.890	2.761		19.24	29.464		32.59	84.540
2 —	6.283	3.141	6¼ –	19.63	30.679	10½—	32.98	86.590
	6.675	3.546		20.02	31.919		33.37	88.664
2¼ –	7.068	3.976	6½—	20.42	33.183	10¾ –	33.77	90.762
	7.461	4.430		20.81	34.471		34.16	92.885
2½—	7.854	4.908	6¾ –	21.20	35 784	11 —	34.55	95.033
	8.246	5.411		21.57	37.122		34.95	97.205
2¾ –	8.639	5.939	7 —	21.99	38.484	11¼ –	35.34	99.402
	9.032	6.491		22.38	39.871		35.73	101.62
3 —	9.424	7.068	7¼ –	22.77	41.282	11½—	36.12	103.86
	9.817	7.669		23.16	42.718		36.52	106.13

THIS TABLE IS TAKEN FROM
"Hamilton's Useful Information for Railway Men,"
Published by D. VAN NOSTRAND,
For the Ramapo Wheel and Foundry Co.

Steamboat Forgings.

De Laney & Co.	Buffalo, N. Y.	See Page 98
Pittsburgh Forge & Iron Co.	Pittsburgh, Pa.	" 174

Steam Elevators.

Crane Bros. Manufacturing Co.	Chicago, Ill.	See Page 230

Steam Engines.

A. Carr	New York	See Page 244
Babcock & Wilcox	" "	" 270
Bolen, Crane & Co.	Newark, N. J.	" 62
Buffalo Machinery Agency	Buffalo, N. Y.	" 160
C. Henry Hall & Co.	New York	" 4
Crane Bros. Manufacturing Co.	Chicago, Ill.	" 230
Erie City Iron Works	Erie, Pa.	" 170
Isaac H. Shearman	Philadelphia, Pa.	" 80
J. B. Waring	New York	" 24
J. S. Mundy	Newark, N. J.	" 266
Miller & Smith	New York	" 112
New York Safety Steam Power Co.	" "	" 270
Phelps & Sanger	Cleveland, Ohio	" 120
Porter, Bell & Co.	Pittsburgh, Pa.	" 128
Rand & Waring Drill & Compressor Co.	New York	See First Page
R. L. Howard & Son	Buffalo, N. Y.	See Page 40
Snyder Bros.	Williamsport, Pa.	" 46
Wiard Locomotive Attachment Co.—(Anti-Explosive)	New York	" 110

Circumferences and Areas of Circles—*Continued.*

Diam.	Circ.	Area.	Diam.	Circ.	Area.	Diam.	Circ.	Area.
11¾ —	36.91	108.43	16 —	50.26	201.06	20¼ —	63.61	322.06
	37.30	110.75		50.65	204.21		64.01	326.05
12 —	37.69	113.09	16¼ —	51.05	207.39	20½ —	64.40	330.06
	38.09	115.46		51.44	210.59		64.79	334.10
12¼ —	38.48	117.85	16½ —	51.83	213.82	20¾ —	65.18	338.16
	38.87	120.27		52.22	217.07		65.58	342.25
12½ —	39.27	122.71	16¾ —	52.62	220.35	21 —	65.97	346.36
	39.66	125.18		53.01	223.65		66.36	350.49
12¾ —	40.05	127.67	17 —	53.40	226.98	21¼ —	66.75	354.65
	40.44	130.19		53.79	230.33		67.15	358.84
13 —	40.84	132.73	17¼ —	54.19	233.70	21½ —	67.54	363.05
	41.23	135.29		54.58	237.10		67.93	367.28
13¼ —	41.62	137.88	17½ —	54.97	240.52	21¾ —	68.32	371.54
	42.01	140.50		55.37	243.97		68.72	375.82
13½ —	42.41	143.13	17¾ —	55.76	247.45	22 —	69.11	380.13
	42.80	145.80		56.16	250.94		69.50	384.46
13¾ —	43.19	148.48	18 —	56.54	254.46	22¼ —	69.90	388.82
	43.58	151.20		56.94	258.01		70.29	393.20
14 —	43.98	153.93	18¼ —	57.33	261.58	22½ —	70.68	397.60
	44.37	156.69		57.72	265.18		71.07	402.03
14¼ —	44.76	159.48	18½ —	58.11	268.80	22¾ —	71.47	406.49
	45.16	162.29		58.51	272.44		71.86	410.97
14½ —	45 55	165.13	18¾ —	58.90	276.11	23 —	72.25	415.47
	45.94	167.98		59.29	279.81		72.64	420.00
14¾ —	46.33	170.87	19 —	59.69	283.52	23¼ —	73.04	424.55
	46.73	173.78		60.08	287.27		73.43	429.13
15 —	47.12	176.71	19¼ —	60.47	291.03	23½ —	73.82	433.73
	47.51	179.67		60.86	294.83		74.21	438.30
15¼ —	47.90	182.65	19½ —	61.26	298.64	23¾ —	74.61	443.01
	48.30	185.66		61.65	302.48		75.	447.69
15½ —	48.69	188.69	19¾ —	62.04	306.35	24 —	75.39	452.39
	49.08	191.74		62.43	310.24		75.79	457.11
15¾ —	49.48	194.82	20 —	62.83	314.16	24¼ —	76.18	461.86
	49.87	197.93		63.22	318.09		76.57	466.63

This table is taken from
"Hamilton's Useful Information for Railway Men,"
Published by D. Van Nostrand,
For the Ramapo Wheel and Foundry Co.

Steam Boilers.

A. Carr	New York	See Page 244
Babcock & Wilcox	" "	" 270
Buffalo Machinery Agency	Buffalo, N. Y.	" 160
C. Henry Hall & Co	New York	" 4
Erie City Iron Works	Erie, Pa	" 170
Isaac H. Shearman	Philadelphia, Pa	" 80
J. B. Waring	New York	" 24
New York Safety Steam Power Co	" "	" 270
Wiard Locomotive Attachment Co.—(Anti-Explosive)	" "	" 110

Steam Fire Engines.

C. Henry Hall & Co	New York	See Page 4
Crane Bros. Manufacturing Co	Chicago, Ill	" 230
George M. Allerton	New York	" 142

Steam Gauges.

A. Carr	New York	See Page 244
A. Fulton's Son & Co	Pittsburgh, Pa	" 116
Buffalo Steam Gauge Co	Buffalo, N. Y	" 170
J. G. Knapp Manufacturing Co	New York	" 16
J. J. Walworth	Chicago, Ill	" 124
L. G. Tillotson & Co	New York	" 94
McNab & Harlin Manufacturing Co	" "	" 220
Phelps & Sanger	Cleveland, Ohio	" 120
Radley & McAlister Manufacturing Co	New York	" 250
T. B. Bickerton & Co	Philadelphia, Pa	" 138
Vose, Dinsmore & Co	New York	" 60

CIRCUMFERENCES AND AREAS OF CIRCLES—*Continued.*

Diam.	Circ.	Area.	Diam.	Circ.	Area.	Diam.	Circ.	Area.
24½ —	76.96	471.43	28¾ -	90.32	649.18	33 —	103.6	855.30
	77.36	476.25		90.71	654.83		104.	861.79
24¾ -	77.75	481.10	29 —	91.10	660.52	33¼ -	104.4	868.30
	78.14	485.97		91.49	666.22		104.8	874.84
25 —	78.54	490.87	29¼ -	91.89	671.95	33½ —	105.2	881.41
	78.93	495.79		92.28	677.71		105.6	888.00
25¼ -	79.32	500.74	29½ —	92.67	683.49	33¾ -	106.	894.61
	79.71	505.71		93.06	689.29		106.4	901.25
25½ —	80.10	510.70	29¾ -	93.46	695.12	34 —	106.8	907.92
	80.50	515.72		93.85	700.98		107.2	914.61
25¾ -	80.89	520.70	30 —	94.24	706.86	34¼ -	107.5	921.32
	81.28	525.83		94.64	712.76		107.9	928.06
26 —	81.68	530.93	30¼ -	95.03	718.69	34½ —	108.3	934.82
	82.07	536.04		95.42	724.64		108.7	941.60
26¼ -	82.46	541.18	30½ —	95.81	730.61	34¾ -	109.1	948.41
	82.85	546.35		96.21	736.61		109.5	955.25
26½ —	83.25	551.54	30¾ -	96.60	742.64	35 —	109.9	962.11
	83.64	556.76		96.99	748.69		110.3	968.99
26¾ -	84.03	562.00	31 —	97.38	754.76	35¼ -	110.7	975.90
	84.43	567.26		97.78	760.86		111.1	982.84
27 —	84.82	572.55	31¼ -	98.17	766.99	35½ —	111.5	989.80
	85.21	577.87		98.56	773.14		111.9	996.78
27¼ -	85.60	583.20	31½ —	98.96	779.31	35¾ -	112.3	1003.7
	86.	588.57		99.35	785.51		112.7	1010.8
27½ —	86.39	593.95	31¾ -	99.74	791.73	36 —	113.	1017.8
	86.78	599.37		100.1	797.97		113.4	1024.9
27¾ -	87.17	604.80	32 —	100.5	804.24	36¼ -	113.8	1032.0
	87.57	610.26		100.9	810.54		114.2	1039.1
28 —	87.96	615.75	32¼ -	101.3	816.86	36½ —	114.6	1046.3
	88.35	621.26		101.7	823.21		115.	1053.5
28¼ -	88.75	626.79	32½ —	102.1	829.57	36¾ -	115.4	1060.7
	89.14	632.35		102.4	835.97		115.8	1067.9
28½ —	89.53	637.94	32¾ -	102.8	842.39	37 —	116.2	1075.2
	89.92	643.54		103.2	848.83		116.6	1082.4

THIS TABLE IS TAKEN FROM
"Hamilton's Useful Information for Railway Men,"
Published by D. VAN NOSTRAND,
For the Ramapo Wheel and Foundry Co.

CIRCUMFERENCES AND AREAS OF CIRCLES—*Continued.*

Diam.	Circ.	Area.	Diam.	Circ.	Area.	Diam.	Circ.	Area.
37¼ –	117.	1089.7	41½ —	130.3	1352.6	45¾ –	143.7	1643.8
	117.4	1097.1		130.7	1360.8		144.1	1652.8
37½ —	117.8	1104.4	41¾ –	131.1	1369.0	46 —	144.5	1661.9
	118.2	1111.8		131.5	1377.2		144.9	1670.9
37¾ –	118.6	1119.2	42 —	131.9	1385.4	46¼ –	145 2	1680.0
	118.9	1126.6		132.3	1393.7		145.6	1689.1
38 —	119.3	1134.1	42¼ –	132.7	1401.9	46½ —	146.	1698.2
	119.7	1141.5		133.1	1410.2		146.4	1707.3
38¼ –	120.1	1149.0	42½ —	133.5	1418.6	46¾ –	146.8	1716.5
	120 5	1156.6		133.9	1426.9		147.2	1725.7
38½ —	120.9	1164.1	42¾ –	134.3	1435.3	47 —	147.6	1734.9
	121.3	1171.7		134.6	1443.7		148.	1744.1
38¾ –	121.7	1179.3	43 —	135.	1452.2	47¼ –	148.4	1753.4
	122.1	1186.9		135.4	1460.6		148.8	1762.7
39 —	122.5	1194.5	43¼ –	135.8	1469.1	47½ —	149.2	1772.0
	122.9	1202.2		136.2	1477.6		149.6	1781.3
39¼ –	123.3	1209.9	43½ —	136.6	1486.1	47¾ –	150.	1790.7
	123.7	1217.6		137.	1494.7		150.4	1800.1
39½ —	124.	1225.4	43¾ –	137.4	1503.3	48 —	150.7	1809.5
	124.4	1233.1		137.8	1511.9		150.1	1818.9
39¾ –	124.8	1240.9	44 —	138.2	1520.5	48¼ –	151.5	1828.4
	125.2	1248.7		138.6	1529.1		151.9	1837.9
40 —	125.6	1256.6	44¼ –	139.	1537.8	48½ —	152.3	1847.4
	126.	1264.5		139.4	1546.5		152.7	1856.9
40¼ –	126.4	1272.3	44½ —	139.8	1555.2	48¾ –	153.1	1866.5
	126.8	1280.3		140.1	1564.0		153.5	1876.1
40½ —	127.2	1288.2	44¾ –	140.5	1572.8	49 —	153.9	1885.7
	127.6	1296.2		140.9	1581.6		154.3	1895.3
40¾ –	128.	1304.2	45 —	141.3	1590.4	49¼ –	154.7	1905.0
	128.4	1312.2		141.7	1599.2		155.1	1914.7
41 —	128.8	1320.2	45¼ –	142.1	1608.1	49½ —	155.5	1924.4
	129.1	1328.3		142.5	1617.0		155.9	1934.1
41¼ –	129.5	1336.4	45½ —	142.9	1625.9	49¾ –	156.2	1943.9
	129.9	1344.5		143.3	1634.9		156.6	1953.6

THIS TABLE IS TAKEN FROM
"Hamilton's Useful Information for Railway Men,"
Published by D. VAN NOSTRAND,
For the Ramapo Wheel and Foundry Co.

MANHATTAN FIRE BRICK

AND

ENAMELED CLAY RETORT WORKS,

MAURER & WEBER, - - - Proprietors,

Office, 633 East 15th St., New York.

LOCOMOTIVE BLOCKS

GAS AND SUGAR HOUSE RETORTS,

Fire Brick, Tiles, &c.,

OF ALL SHAPES AND SIZES.

SUPERIOR FIRE CLAY,

From our own Clay Beds near Perth Amboy, New Jersey, by the Ton or Cargo.

Steam Pipe.

A. Carr....New York....See Page 244
Evans, Dalzell & Co....Pittsburgh, Pa.... " 58
J. J. Walworth....Chicago, Ill.... " 124
J. T. Ryerson.... " " .... " 180
Phelps & Sanger....Cleveland, Ohio.... " 120
Van Tuyl Manufacturing Co....New York.... " 242
Winans & Co....New York.... " 106
Wm. Graff & Co....Pittsburgh, Pa.... " 174

Steam Pipe Covering.

H. W. Johns....New York....See Page 224
Van Tuyl Manufacturing Co.... " " .... " 242

Steamship Lines.

Inman Line....New York....See Page 14

Steam Ship Supplies.

Wm. Green & Co....Wilmington, Del....See Page 248

Steam Warming Apparatus.

Crane Bros. Manufacturing Co....Chicago, Ill....See Page 230

Steam Whistles.

J. J. Walworth....Chicago, Ill....See Page 124
McNab & Harlin Manufacturing Co....New York.... " 220
Pancoast & Maule....Philadelphia, Pa.... " 26
Vose, Dinsmore & Co....New York.... " 60

CIRCUMFERENCES AND AREAS OF CIRCLES—*Continued.*

Diam.	Circ.	Area.	Diam.	Circ.	Area.	Diam.	Circ.	Area.
50 —	157.	1963.5	54¼ –	170.4	2311.4	58½—	183.7	2687.8
	157.4	1973.3		170.8	2322.1		184.1	2699.3
50¼ –	157.8	1983.1	54½—	171.2	2332.8	58¾ –	184.5	2710.8
	158.2	1993.0		171.6	2343.5		184.9	2722.4
50½—	158.6	2002.9	54¾ –	172.	2354.2	59 —	185.3	2733.9
	159.	2012.8		172.3	2365.0		185.7	2745.5
50¾ –	159.4	2022.8	55 —	172.7	2375.8	59¼ –	186.1	2757.1
	159.8	2032.8		173.1	2386.6		186.5	2768.8
51 —	160.2	2042.8	55¼ –	173.5	2397.4	59½—	186.9	2780.5
	160.6	2052.8		173.9	2408.3		187.3	2792.2
51¼ –	161.	2062.9	55½—	174.3	2419.2	59¾ –	187.7	2803.9
	161.3	2072.9		174.7	2430.1		188.1	2815.6
51½—	161.7	2083.0	55¾ –	175.1	2441.0	60 —	188.4	2827.4
	162.1	2093.2		175.5	2452.0		188.8	2839.2
51¾ –	162.5	2103.3	56 —	175.9	2463.0	60¼ –	189.2	2851.0
	162.9	2113.5		176.3	2474.0		189.6	2862.8
52 —	163.3	2123.7	56¼ –	176.7	2485.0	60½—	190.	2874.7
	163.7	2133.9		177.1	2496.1		190.4	2886.6
52¼ –	164.1	2144.1	56½—	177.5	2507.1	60¾ –	190.8	2898.5
	164.5	2154.4		177.8	2518.2		191.2	2910.5
52½—	164.9	2164.7	56¾ –	178.2	2529.4	61 —	191.6	2922.4
	165.3	2175.0		178.6	2540.5		192.	2934.4
52¾ –	165.7	2185.4	57 —	179.	2551.7	61¼ –	192.4	2946.4
	166.1	2195.7		179.4	2562.9		192.8	2958.5
53 —	166.5	2206.1	57¼ –	179.8	2574.1	61½—	193.2	2970.5
	166.8	2216.6		180.2	2585.4		193.6	2982.6
53¼ –	167.2	2227.0	57½—	180.6	2596.7	61¾ –	193.9	2994.7
	167.6	2237.5		181.	2608.0		194.3	3006.9
53½—	168.	2248.0	57¾ –	181.4	2619.3	62 —	194.7	3019.0
	168.4	2258.5		181.8	2630.7		195.1	3031.2
53¾ –	168 8	2269.0	58 —	182.2	2642.0	62¼ –	195.5	3043.4
	169.2	2279.6		182.6	2653.4		195.9	3055.7
54 —	169.6	2290.2	58¼ –	182.9	2664.9	62½—	196.3	3067.9
	170.	2300.8		183.3	2676.3		196.7	3080.2

HARDEN'S ANTI-FRICTION GLASS BEARING
PATENTED JANUARY 14TH 1868

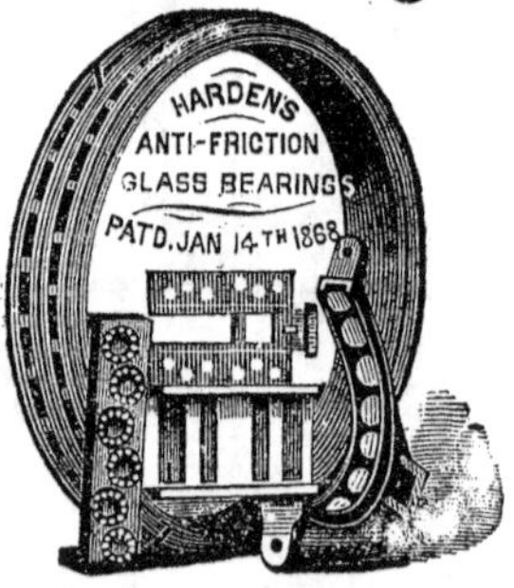
HARDEN'S
ANTI-FRICTION
GLASS BEARINGS
PAT'D. JAN 14TH 1868

Steel.

Anderson & Woods	Pittsburgh, Pa.	See Page 150
Atlantic Steel Works	New York	" 212
Chas. W. Matthews	Philadelphia, Pa.	" 38
England & Bindley	Pittsburgh, Pa.	" 162
Posts & Kalkman	New York	" 266
Reese, Graff & Woods	Pittsburgh, Pa.	" 68
Tyng & Co.	New York	" 260
Wm. Green & Co.	Wilmington, Del.	" 248

See also, Steel Importers and Steel Manufacturers.

Steel Bells.

A. Fulton's Son & Co.	Pittsburgh, Pa.	See Page 116
Hermann Boker & Co.	New York	See Pages 152, 154

Steel Car Springs.

Chicago Spring Works	Chicago, Ill.	See Page 200
Greene & Randolph	New York	" 54
Vose, Dinsmore & Co.	" "	" 60

Steel Forgings.

Anderson & Woods	Pittsburgh, Pa.	See Page 150
Atlantic Steel Works	New York	" 212
De Laney & Co.	Buffalo, N. Y.	" 98
Jno. A. Griswold & Co.	Troy, N. Y.	" 50

Steel Frogs.

Anderson & Woods	Pittsburgh, Pa.	See Page 150
Atlantic Steel Works	New York	" 212

CIRCUMFERENCES AND AREAS OF CIRCLES—*Continued.*

Diam.	Circ.	Area.	Diam.	Circ.	Area.	Diam.	Circ.	Area.
62¾ –	197.1	3092.5	67 —	210.4	3525.6	71¼ –	223.8	3987.1
	197.5	3104.8		210.8	3538.8		224.2	4001.1
63 —	197.9	3117.2	67¼ –	211.2	3552.0	71½—	224.6	4015.1
	198.3	3129.6		211.6	3565.2		225.	4029.2
63¼ –	198.7	3142.0	67½—	212.	3578.4	71¾ –	225.4	4043.2
	199.	3154.4		212.4	3591.7		225.8	4067.3
63½—	199.4	3166.9	67¾ –	212.8	3605.0	72 —	226.1	4071.5
	199.8	3179.4		213.2	3618.3		226.5	4085.6
63¾ –	200.2	3191.9	68 —	213.6	3631.6	72¼ –	226.9	4099.8
	200.6	3204.4		214.	3645.0		227.3	4114.0
64 —	201.	3216.9	68¼ –	214.4	3658.4	72½—	227.7	4128.2
	201.4	3229.5		214.8	3671.8		228.1	4142.5
64¼ –	201.8	3242.1	68½—	215.1	3685.2	72¾ –	228.5	4156.7
	202.2	3254.8		215.5	3698.7		228.9	4171.0
64½—	202.6	3267.4	68¾ –	215.9	3712.2	73 —	229.3	4185.3
	203.	3280.1		216.3	3725.7		229.7	4199.7
64¾ –	203.4	3292.8	69 —	216.7	3739.2	73¼ –	230.1	4214.1
	203.8	3205.5		217.1	3752.8		230.5	4228.5
65 —	204.2	3318.3	69¼ –	217.5	3766.4	73½—	230.9	4242.9
	204.5	3331.0		217.9	3780.0		231.3	4257.3
65¼ –	204.9	3343.8	69½—	218.3	3793.6	73¾ –	231.6	4271.8
	205.3	3356.7		218.7	3807.3		232.	4286.3
65½—	205.7	3369.5	69¾ –	219.1	3821.0	74 —	232.4	4300.8
	206.1	3382.4		219.5	3834.7		232.8	4315.3
65¾ –	206.5	3395.3	70 —	219.9	3848.4	74¼ –	233.2	4329.9
	206.9	3408.2		220.3	3862.2		233.6	4344.5
66 —	207.3	3421.2	70¼ –	220.6	3875.9	74½—	234.	4359.1
	207.7	3434.1		221.	3889.8		234.4	4373.8
66¼ –	208.1	3447.1	70½—	221.4	3903.6	74¾ –	234.8	4388.4
	208.5	3460.1		221.8	3917.4		235.2	4403.1
66½—	208.9	3473.2	70¾ –	222.2	3931.3	75 —	235.6	4417.8
	209.3	3486.3		222.6	3945.2		236.	4432.6
66¾ –	209.7	3499.3	71 —	223.	3959.2	75¼ –	236.4	4447.3
	210.	3512.5		223.4	3973.1		236.7	4462.1

THIS TABLE IS TAKEN FROM
"Hamilton's Useful Information for Railway Men,"
Published by D. VAN NOSTRAND,
For the Ramapo Wheel and Foundry Co.

CIRCUMFERENCES AND AREAS OF CIRCLES—*Continued.*

Diam.	Circ.	Area.	Diam.	Circ.	Area.	Diam.	Circ.	Area.
75½—	237.1	4476.9	79¾ –	250.5	4995.1	84 —	263.8	5541.7
	237.5	4491.8		250.9	5010.8		264.2	5558.2
75¾ –	237.9	4506.6	80 —	251.3	5026.5	84¼ –	264.6	5574.8
	238.3	4521.5		251.7	5042.2		265.	5591.3
76 —	238.7	4536.4	80¼ –	252.1	5058.0	84½—	265.4	5607 9
	239.1	4551.4		252.5	5073.7		265.8	5624.5
76¼ –	239.5	4566.3	80½—	252.8	5089.5	84¾ –	266.2	5641.1
	239.9	4581.3		253.2	5105 4		266.6	5657.8
76½—	240.3	4596.3	80¾ –	253.6	5121.2	85 —	267.	5674.5
	240.7	4611.3		254.	5137.1		267.4	5691.2
76¾ –	241.1	4626.4	81 —	254.4	5153.0	85¼ –	267.8	5707.9
	241.5	4641.5		254.8	5168.9		268.2	5724.6
77 —	241.9	4656.6	81¼ –	255.2	5184.8	85½—	269.6	5741.4
	242.2	4671.7		255.6	5200.8		268.9	5758.2
77¼ –	242.6	4686.9	81½—	256.	5216.8	85¾ –	269.3	5775.0
	243.	4702.1		256.4	5232.8		269.7	5791.9
77½—	243.4	4717.3	81¾ –	256.8	5248.8	86 —	270.1	5808.8
	243.8	4732.5		257.2	5264.9		270.5	5825.7
77¾ –	244.2	4747.7	82 —	257.6	5281.0	86¼ –	270.9	5842.6
	244.6	4763.0		258.	5297.1		271.3	5859.5
78 —	245.	4778.3	82¼ –	258.3	5313.2	86½—	271.7	5876.5
	245.4	4793.7		258.7	5329.4		272.1	5893.5
78¼ –	245.8	4809.0	82½—	259.1	5345.6	86¾ –	272.5	5910.5
	246.2	4824.4		259.5	5361.8		272.9	5927.6
78½—	246.6	4839.8	82¾ –	259.9	5378.0	87 —	273.3	5944.6
	247.	4855.2		260.3	5394.3		273.7	5961.7
78¾ –	247.4	4870.7	83 —	260.7	5410.6	87¼ –	274.1	5978.9
	247.7	4886.1		261.1	5426.9		274.4	5996.0
79 —	248.1	4901.6	83¼ –	261.5	5443.2	87½—	274.8	6013.2
	248.5	4917.2		261.9	5459.6		275.2	6030.4
79¼ –	248.9	4932.7	83½—	262.3	5476.0	87¾ –	275.6	6047.6
	249.3	4948.3		262.7	5492.4		276.	6064.8
79½—	249.7	4963.9	83¾ –	263.1	5508.8	88 —	276.4	6082.1
	250.1	4979.5		263.5	5525.3		276.8	6099.4

THIS TABLE IS TAKEN FROM
"Hamilton's Useful Information for Railway Men,"
Published by D. VAN NOSTRAND,
For the Ramapo Wheel and Foundry Co.

JAY & COOK,

MANUFACTURERS OF

ZINC GLAZIER'S POINTS.

SHARPS.

TRIANGLES.

No. 43 Centre Street,

NEW YORK.

CIRCUMFERENCES AND AREAS OF CIRCLES—*Continued.*

Diam.	Circ.	Area.	Diam.	Circ.	Area.	Diam.	Circ.	Area.
88¼	277.2	6116.7	92½	290.5	6720.0	96¾	303.9	7351.7
	277.6	6134.0		290.9	6738.2		304.3	7370.7
88½	278.	6151.4	92¾	291.3	6756.4	97	304.7	7389.8
	278.4	6168.8		291.7	6776.4		305.1	7408.8
88¾	278.8	6186.2	93	292.1	6792.9	97¼	305.5	7427.9
	279.2	6203.6		292.5	6811.1		305.9	7447.0
89	279.6	6221.1	93¼	292.9	6829.4	97½	306.3	7466.2
	279.9	6238.6		293.3	6847.8		306.6	7485.3
89¼	280.3	6256.1	93½	293.7	6866.1	97¾	307.	7504.5
	280.7	6273.6		294.1	6884.5		307.4	7523.7
89½	281.1	6291.2	93¾	294.5	6902.9	98	307.8	7542.9
	281.5	6308.8		294.9	6921.3		308.2	7562.2
89¾	281.9	6326.4	94	295.3	6939.7	98¼	308.6	7581.5
	282.3	6344.0		295.7	6958.2		309.0	7600.8
90	282.7	6361.7	94¼	296.	6976.7	98½	309.4	7620.1
	283.1	6379.4		296.4	6995.2		309.8	7639.4
90¼	283.5	6397.1	94½	296.8	7013.8	98¾	310.2	7658.8
	283.9	6414.8		297.2	7032.3		310.6	7678.2
90½	284.3	6432.6	94¾	297.6	7050.9	99	311.0	7697.7
	284.7	6450.4		298.	7069.5		311.4	7717.1
90¾	285.1	6468.2	95	298.4	7088.2	99¼	311.8	7736.6
	285.4	6486.0		298.8	7106.9		312.1	7756.1
91	285.8	6503.8	95¼	299.2	7125.5	99½	312.5	7775.6
	286.2	6521.7		299.6	7144.3		312.9	7795.2
91¼	286.6	6539.6	95½	300.	7163.0	99¾	313.3	7814.7
	287.	6557.6		300.4	7181.8		313.7	7834.3
91½	287.4	6575.5	95¾	300.8	7200.5	100	314.1	7853.6
	287.8	6593.5		301.2	7219.4		314.5	7853.9
91¾	288.2	6611.5	96	301.5	7238.2	100¼	314.9	7893.3
	288.6	6629.5		301.9	7257.1		315.3	7913.1
92	289.	6647.6	96¼	302.3	7275.9	100½	315.7	7932.7
	289.4	6665.7		302.7	7294.9		316.0	7942.4
92¼	289.8	6683.8	96½	303.1	7313.8	100¾	316.4	7972.2
	290.2	6701.9		303.5	7332.8		316.8	7991.9

THIS TABLE IS TAKEN FROM
"Hamilton's Useful Information for Railway Men,"
Published by D. VAN NOSTRAND,
For the Ramapo Wheel and Foundry Co.

Squares, Cubes, Square Roots, and Cube Roots.

No.	Square.	Cube.	Square Root.	Cube Root.	No.	Square.	Cube.	Square Root.	Cube Root.
1	1	1	1.0	1.0	31	961	29791	5.567	3.141
2	4	8	1.414	1.259	32	1024	32768	5.656	3.174
3	9	27	1.732	1.442	33	1089	35937	5.744	3.207
4	16	64	2.0	1.587	34	1156	39304	5.830	3.239
5	25	125	2.230	1.709	35	1225	42875	5.916	3.271
6	36	216	2.449	1.817	36	1296	46656	6.0	3.301
7	49	343	2.645	1.912	37	1369	50653	6.082	3.332
8	64	512	2.828	2.0	38	1444	54872	6.164	3.361
9	81	729	3.0	2.080	39	1521	59319	6.245	3.391
10	100	1000	3.162	2.154	40	1600	64000	6.324	3.419
11	121	1331	3.316	2.223	41	1681	68921	6.403	3.448
12	144	1728	3.464	2.289	42	1764	74088	6.480	3.476
13	169	2197	3.605	2.351	43	1849	79507	6.557	3.503
14	196	2744	3.741	2.410	44	1936	85184	6.633	3.530
15	225	3375	3.872	2.466	45	2025	91125	6.708	3.556
16	256	4096	4.0	2.519	46	2116	97336	6.782	3.583
17	289	4913	4.123	2.571	47	2209	103823	6.855	3.608
18	324	5832	4.242	2.620	48	2304	110592	6.928	3.634
19	361	6859	4.358	2.668	49	2401	117649	7.0	3.659
20	400	8000	4.472	2.714	50	2500	125000	7.071	3.684
21	441	9261	4.582	2.758	51	2601	132651	7.141	3.708
22	484	10648	4.690	2.802	52	2704	140608	7.211	3.732
23	529	12167	4.795	2.843	53	2809	148877	7.280	3.756
24	576	13824	4.898	2.884	54	2916	157464	7.348	3.779
25	625	15625	5.0	2.924	55	3025	166375	7.416	3.802
26	676	17576	5.099	2.962	56	3136	175616	7.483	3.825
27	729	19683	5.196	3.0	57	3249	185193	7.549	3.848
28	784	21952	5.291	3.036	58	3364	195112	7.615	3.870
29	841	24389	5.385	3.072	59	3481	205379	7.681	3.893
30	900	27000	5.477	3.107	60	3600	216000	7.745	3.914

THIS TABLE IS TAKEN FROM
"Hamilton's Useful Information for Railway Men,"
Published by D. Van Nostrand,
For the Ramapo Wheel and Foundry Co.

Tallow Cans and Cups.

J. G. Knapp Manufacturing Co........................New York........................See Page 16
Pancoast & Maule..Philadelphia, Pa............ " 26

Tallow Lubricator.

Isaac U. Coles..New York.....................See Page 20

Tank Iron.

J. T. Ryerson...Chicago, Ill...................See Page 180

Tanks.

Joseph Churchyard ....Buffalo, N. Y.................See Page 160

Tank Valves.

Hart, Ball & Hart..Buffalo, N. Y..................See Page 136
Skinner & Gifford Manufacturing Co...............Dunkirk, N. Y................. " 140

Taps.

Champlin & Rogers..Chicago, Ill.....................See Page 192
Providence Tool Co.; H. B. Newhall, Agent.....New York........................ " 166
The Worcester Machine Screw Co.; H. B. Newhall, Agent.................................... " " .... " 166

SQUARES, CUBES, SQUARE ROOTS, AND CUBE ROOTS—*Continued.*

No.	Square.	Cube.	Square Root.	Cube Root.	No.	Square.	Cube.	Square Root.	Cube Root.
61	3721	226981	7.810	3.936	91	8281	753571	9.539	4.497
62	3844	238328	7.874	3.957	92	8464	778688	9.591	4.514
63	3969	250047	7.937	3.979	93	8649	804357	9.643	4.530
64	4096	262144	8.0	4.0	94	8836	830584	9.695	4.546
65	4225	274625	8.062	4.020	95	9025	857375	9.746	4.562
66	4356	287496	8.124	4.041	96	9216	884736	9.797	4.578
67	4489	300763	8.185	4.061	97	9409	912673	9.848	4.594
68	4624	314432	8.246	4.081	98	9604	941192	9.899	4.610
69	4761	328509	8.306	4.101	99	9801	970299	9.949	4.626
70	4900	343000	8.366	4.121	100	10000	1000000	10.0	4.641
71	5041	357911	8.426	4.140	101	10201	1030301	10.049	4.657
72	5184	373248	8.485	4.160	102	10404	1061208	10.099	4.672
73	5329	389017	8.544	4.179	103	10609	1092727	10.148	4.687
74	5476	405224	8.602	4.198	104	10816	1124864	10.198	4.702
75	5625	421875	8.660	4.217	105	11025	1157625	10.246	4.717
76	5776	438976	8.717	4.235	106	11236	1191016	10.295	4.732
77	5929	456533	8.774	4.254	107	11449	1225043	10.344	4.747
78	6084	474552	8.831	4.272	108	11664	1259712	10.392	4.762
79	6241	493039	8.888	4.290	109	11881	1295029	10.440	4.776
80	6400	512000	8.944	4.308	110	12100	1331000	10.488	4.791
81	6561	531441	9.0	4.326	111	12321	1367631	10.535	4.805
82	6724	551368	9.055	4.344	112	12544	1404928	10.583	4.820
83	6889	571787	9.110	4.362	113	12769	1442897	10.630	4.834
84	7056	592704	9.165	4.379	114	12996	1481544	10.677	4.848
85	7225	614125	9.219	4.396	115	13225	1520875	10.723	4.862
86	7396	636056	9.273	4.414	116	13456	1560896	10.770	4.877
87	7569	658503	9.327	4.431	117	13689	1601613	10.816	4.890
88	7744	681472	9.380	4.447	118	13924	1643032	10.862	4.904
89	7921	704969	9.433	4.464	119	14161	1685159	10.908	4.918
90	8100	729000	9.486	4.481	120	14400	1728000	10.954	4.932

THIS TABLE IS TAKEN FROM
"Hamilton's Useful Information for Railway Men,"
Published by D. VAN NOSTRAND,
For the Ramapo Wheel and Foundry Co.

Telegraph Apparatus.

Charles T. Chester....New York....See Page 212
F. L. Pope & Co.... " " .... " 184
Geo. H. Crain & Co....Chicago, Ill.... " 124
Geo. H. Bliss & Co.... " " .... " 164
Leclanche Battery Co....New York.... " 96
L. G. Tillotson & Co.... " " .... " 94
M. A. Buell....Cleveland, Ohio.... " 46
Merchants Manufacturing & Construction Co....New York.... " 18
Mathiesen & Schrader.... " " .... " 82
Towle & Unger Manufacturing Co.... " " .... " 186
Vose, Dinsmore & Co.... " " .... " 60

Telegraph Batteries.

See Telegraph Apparatus.

Telegraph Builders.

Charles T. Chester....New York....See Page 212
Merchants Manufacturing & Construction Co.... " " .... " 18

Telegraph Instruments and Fixtures.

See Telegraph Apparatus.

Telegraph Material.

Charles T. Chester....New York....See Page 212
F. L. Pope & Co.... " " .... " 184
Geo. H Bliss & Co....Chicago, Ill.... " 164
Leclanche Battery Co....New York.... " 96
L. G. Tillotson & Co.... " " .... " 94
M. A. Buell....Cleveland, Ohio.... " 46
Merchants Manufacturing & Construction Co. New York.... " 18

SQUARES, CUBES, SQUARE ROOTS, AND CUBE ROOTS—*Continued.*

No.	Square.	Cube.	Square Root.	Cube Root.	No.	Square.	Cube.	Square Root.	Cube Root.
121	14641	1771561	11.0	4.946	151	22801	3442951	12.288	5.325
122	14884	1815848	11.045	4.959	152	23104	3511008	12.328	5.336
123	15129	1860867	11.090	4.973	153	23409	3581577	12.369	5.348
124	15376	1906624	11.135	4.986	154	23716	3652264	12.409	5.360
125	15625	1953125	11.180	5.0	155	24025	3723875	12.449	5.371
126	15876	2000376	11.224	5.013	156	24336	3796416	12.490	5.383
127	16129	2048383	11.269	5.026	157	24649	3869893	12.529	5.394
128	16384	2097152	11.313	5.039	158	24964	3944312	12.569	5.406
129	16641	2146689	11.357	5.052	159	25281	4019679	12.609	5.417
130	16900	2197000	11.401	5.065	160	25600	4096000	12.649	5.428
131	17161	2248091	11.445	5.078	161	25921	4173281	12.688	5.440
132	17424	2299968	11.489	5.091	162	26244	4251528	12.727	5.451
133	17689	2352637	11.532	5.104	163	26569	4330747	12.767	5.462
134	17956	2406104	11.575	5.117	164	26896	4410944	12.806	5.473
135	18225	2460375	11.618	5.129	165	27225	4492125	12.845	5.484
136	18496	2515456	11.661	5.142	166	27556	4574296	12.884	5.495
137	18769	2571353	11.704	5.155	167	27889	4657463	12.922	5.506
138	19044	2628072	11.747	5.167	168	28224	4741632	12.961	5.517
139	19321	2685619	11.789	5.180	169	28561	4826809	13.0	5.528
140	19600	2744000	11.832	5.192	170	28900	4913000	13.038	5.539
141	19881	2803221	11.874	5.204	171	29241	5000211	13.076	5.550
142	20164	2863288	11.916	5.217	172	29584	5088448	13.114	5.561
143	20449	2924207	11.958	5.229	173	29929	5177717	13.152	5.572
144	20736	2985984	12.0	5.241	174	30276	5268024	13.190	5.582
145	21025	3048625	12.041	5.253	175	30625	5359375	13.228	5.593
146	21316	3112136	12.083	5.265	176	30976	5451776	13.266	5.604
147	21609	3176523	12.124	5.277	177	31329	5545233	13.304	5.614
148	21904	3241792	12.165	5.289	178	31684	5639752	13.341	5.625
149	22201	3307949	12.206	5.301	179	32041	5735339	13.379	5.635
150	22500	3375000	12.247	5.313	180	32400	5832000	13.416	5.646

THIS TABLE IS TAKEN FROM
"Hamilton's Useful Information for Railway Men,"
Published by D. VAN NOSTRAND,
For the Ramapo Wheel and Foundry Co.

Telegraph Poles.

Geo. H. Crain & Co....Chicago, Ill....See Page 124
Geo. H. Bliss & Co.... " " .... " 164

Telegraph Repairers' Tools.

Leclanche Battery Co....New York....See Page 96
See also, Telegraph Material.

Telegraph Wires.

Bishop Gutta Percha Works....New York....See Page 34
Chas. T. Chester.... " " .... " 212
C. Thompson....Philadelphia, Pa.... " 222
Geo. H. Bliss & Co....Chicago, Ill.... " 164
Geo. H. Crain & Co.... " " .... " 124
John Merry & Co....New York.... " 48
Leclanche Battery Co.... " " .... " 96
Marshall Lefferts, Jr.... " " .... " 44
Tyng & Co.... " " .... " 260

Tiles.

Maurer & Weber....New York....See Page 228
Philip Neukumet....Philadelphia, Pa.... " 104
The Star Fire Brick Co....Pittsburgh, Pa.... " 232

Timber.

Joseph Churchyard....Buffalo, N. Y....See Page 160

Tin—Pig.

Holmes & Lissberger....New York....See Page 84
Philadelphia Smelting Co....Philadelphia, Pa.... " 134

SQUARES, CUBES, SQUARE ROOTS, AND CUBE ROOTS—*Continued.*

No.	Square.	Cube.	Square Root.	Cube Root.	No.	Square.	Cube.	Square Root.	Cube Root.
181	32761	5929741	13.453	5.656	211	44521	9393931	14.525	5.953
182	33124	6028568	13.490	5.667	212	44944	9528128	14.560	5.962
183	33489	6128487	13.527	5.677	213	45369	9663597	14.594	5.972
184	33856	6229504	13.564	5.687	214	45796	9800344	14.628	5.981
185	34225	6331625	13.601	5.698	215	46225	9938375	14.662	5.990
186	34596	6434856	13.638	5.708	216	46656	10077696	14.696	6.0
187	34969	6539203	13.674	5.718	217	47089	10218313	14.730	6.009
188	35344	6644672	13.711	5.728	218	47524	10360232	14.764	6.018
189	35721	6751269	13.747	5.738	219	47961	10503459	14.798	6.027
190	36100	6859000	13.784	5.748	220	48400	10648000	14.832	6.036
191	36481	6967871	13.820	5.758	221	48841	10793861	14.866	6.045
192	36864	7077888	13.856	5.769	222	49284	10941048	14.899	6.055
193	37249	7189517	13.892	5.779	223	49729	11089567	14.933	6.064
194	37636	7301384	13.928	5.788	224	50176	11239424	14.966	6.073
195	38025	7414875	13.964	5.798	225	50625	11390625	15.0	6.082
196	38416	7529536	14.0	5.808	226	51076	11543176	15.033	6.099
197	38809	7645373	14.035	5.818	227	51529	11697083	15.066	6.100
198	39204	7762392	14.071	5.828	228	51984	11852352	15.099	6.109
199	39601	7880599	14.106	5.838	229	52441	12008989	15.132	6.118
200	40000	8000000	14.142	5.848	230	52900	12167000	15.165	6.126
201	40401	8120601	14.177	5.857	231	53361	12326391	15.198	6.135
202	40804	8242408	14.212	5.867	232	53824	12487168	15.231	6.144
203	41209	8365427	14.247	5.877	233	54289	12649337	15.264	6.153
204	41616	8489664	14.282	5.886	234	54756	12812904	15.297	6.162
205	42025	8615125	14.317	5.896	235	55225	12977875	15.329	6.171
206	42436	8741816	14.352	5.905	236	55696	13144256	15.362	6.179
207	42849	8869743	14.387	5.915	237	56169	13312053	15.394	6.188
208	43264	8998912	14.422	5.924	238	56644	13481272	15.427	6.197
209	43681	9129329	14.456	5.934	239	57121	13651919	15.459	6.205
210	44100	9261000	14.491	5.943	240	57600	13824000	15.491	6.214

THIS TABLE IS TAKEN FROM
"Hamilton's Useful Information for Railway Men,"
Published by D. VAN NOSTRAND,
For the Ramapo Wheel and Foundry Co.

Tin and Coppersmiths' Solders.

Du Plaine & Reeves..Philadelphia, Pa............See Page 102
Philadelphia Smelting Co.............................. " " .... " 134

Tin—Plate.

Bruce & Cook..New York......................See Page 112
Tyng & Co.. " " .... " 260
Wm. Green & Co...Wilmington, Del........... " 248

Tinsmiths' Tools, &c.

Bruce & Cook..New York......................See Page 112

Tin Ware.

Sidney Shepard & Co..................................Buffalo, N. Y..................See Page 196

T Iron.

Rogers & Co...Chicago, Ill....................See Page 144

Tool Builders.

Bolen, Crane & Co......................................Newark, N. J..................See Page 62

Tool Handles.

G. B. Walbridge...New York......................See Page 130

Tools.

Biddle Manufacturing Co.............................New York......................See Page 12
Hermann Boker & Co.................................. " " ....See Pages 152, 154
Nelson Tool Works...................................... " " ....See Page 30
S. S. Townsend... " " .... " 2

Track Bolts.

Pittsburgh Forge & Iron Co..........................Pittsburgh, Pa...............See Page 174
Pittsburgh Bolt Co...................................... " .... " 118

SQUARES, CUBES, SQUARE ROOTS, AND CUBE ROOTS—*Continued.*

No.	Square.	Cube.	Square Root.	Cube Root.	No.	Square.	Cube.	Square Root.	Cube Root.
241	58081	13997521	15.524	6.233	271	73441	19902511	16.462	6.471
242	58564	14172488	15.556	6.231	272	73984	20123648	16.492	6.479
243	59049	14348907	15.588	6.240	273	74529	20346417	16.522	6.487
244	59536	14526784	15.620	6.248	274	75076	20570824	16.552	6.495
245	60025	14706125	15.652	6.257	275	75625	20796875	16.583	6.502
246	60516	14886936	15.684	6.265	276	76176	21024576	16.613	6.510
247	61009	15069223	15.716	6.274	277	76729	21253933	16.643	6.518
248	61504	15252992	15.748	6.282	278	77284	21484952	16.678	6.526
249	62001	15438249	15.779	6.291	279	77841	21717639	16.703	6.534
250	62500	15625000	15.811	6.299	280	78400	21952000	16.733	6.542
251	63001	15813251	15.842	6.307	281	78961	22188041	16.763	6.549
252	63504	16003008	15.874	6.316	282	79524	22425768	16.792	6.557
253	64009	16194277	15.905	6.324	283	80089	22665187	16.822	6.565
254	64516	16387064	15.937	6.333	284	80656	22906304	16.852	6.573
255	65025	16581375	15.968	6.341	285	81225	23149125	16.881	6.580
256	65536	16777216	16.0	6.349	286	81796	23393656	16.911	6.588
257	66049	16974593	16.031	6.357	287	82369	23639903	16.941	6.596
258	66564	17173512	16.062	6.366	288	82944	23887872	16.970	6.603
259	67081	17373979	16.093	6.374	289	83521	24137569	17.0	6.611
260	67600	17576000	16.124	6.382	290	84100	24389000	17.029	6.619
261	68121	17779581	16.155	6.390	291	84681	24642171	17.058	6.626
262	68644	17984728	16.186	6.398	292	85264	24897088	17.088	6.634
263	69169	18191447	16.217	6.406	293	85849	25153757	17.117	6.641
264	69696	18399744	16.248	6.415	294	86436	25412184	17.146	6.649
265	70225	18609625	16.278	6.423	295	87025	25672375	17.175	6.656
266	70756	18821096	16.309	6.431	296	87616	25934836	17.204	6.664
267	71289	19034163	16.340	6.439	297	88209	26198073	17.233	6.671
268	71824	19248832	16.370	6.447	298	88804	26463592	17.262	6.678
269	72361	19465109	16.401	6.455	299	89401	26730899	17.291	6.686
270	72900	19683000	16.431	6.463	300	90000	27000000	17.320	6.694

THIS TABLE IS TAKEN FROM
"Hamilton's Useful Information for Railway Men,"
Published by D. VAN NOSTRAND,
For the Ramapo Wheel and Foundry Co.

Track Layers' Tools.

Nelson Tool Works..........New York..........See Page 30

Track Supplies.

Greene & Randolph..........New York..........See Page 54

T Rails.

Jones & Laughlins..........Pittsburgh, Pa..........See Page 32
8, 12, 16, 20, 28, 40 lbs. to Yard.

Reese, Graff & Woods..........Pittsburgh, Pa..........See Page 68
12, 16, 20, 28 and 34 lbs. to Yard.

Transits.

F. Eckel..........New York..........See Page 76
Kuebler & Seelhorst..........Philadelphia, Pa.......... " 216
Wm. J. Young & Son.......... " " .... " 138

Transfer Tables.

McNairy & Claflen Manufacturing Co..........Cleveland, Ohio..........See Page 42
Skinner & Gifford Manufacturing Co..........Dunkirk, N. Y.......... " 140
Snyder Bros..........Williamsport, Pa.......... " 46
Wm. Sellers & Co..........Philadelphia, Pa.......... " 64

Transportation Cars.

Orrin L. Gridley..........Buffalo, N. Y..........See Page 196
Sidney Shepard & Co.......... " .... " 196

SQUARES, CUBES, SQUARE ROOTS, AND CUBE ROOTS—*Continued.*

No.	Square.	Cube.	Square Root.	Cube Root.	No.	Square.	Cube.	Square Root.	Cube Root.
301	90601	27270901	17.349	6.701	331	109561	36264691	18.193	6.917
302	91204	27543608	17.378	6.709	332	110224	36594368	18.220	6.924
303	91809	27818127	17.406	6.716	333	110889	36926037	18.248	6.931
304	92416	28094464	17.435	6.723	334	111556	37259704	18.275	6.938
305	93025	28372625	17.464	6.731	335	112225	37595375	18.303	6.945
306	93636	28652616	17.492	6.738	336	112896	37933056	18.330	6.952
307	94249	28934443	17.521	6.745	337	113569	38272753	18.357	6.958
308	94864	29218112	17.549	6.753	338	114244	38614472	18.384	6.965
309	95481	29503609	17.578	6.760	339	114921	38958219	18.411	6.972
310	96100	29791000	17.606	6.767	340	115600	39304000	18.439	6.979
311	96721	30080231	17.635	6.775	341	116281	39651821	18.466	6.986
312	97344	30371328	17.663	6.782	342	116964	40001688	18.493	6.993
313	97969	30664297	17.691	6.789	343	117649	40353607	18.520	7.0
314	98596	30959144	17.720	6.796	344	118336	40707584	18.547	7.006
315	99225	31255875	17.748	6.804	345	119025	41063625	18.574	7.013
316	99856	31554496	17.776	6.811	346	119716	41421736	18.601	7.020
317	100489	31855013	17.804	6.818	347	120409	41781923	18 627	7.027
318	101124	32157432	17.832	6.825	348	121104	42144192	18.654	7.033
319	101761	32461759	17.860	6.832	349	121801	42508549	18.681	7.040
320	102400	32768000	17.888	6.839	350	122500	42875000	18.708	7.047
321	103041	33076161	17.916	6.847	351	123201	43243551	18.734	7.054
322	103684	33386248	17.944	6.854	352	123904	43614208	18.761	7.060
323	104329	33698267	17.972	6.861	353	124609	43986977	18.788	7.067
324	104976	34012224	18.0	6 868	354	125316	44361864	18.814	7.074
325	105625	34328125	18.027	6.875	355	126025	44738875	18.841	7 080
326	106276	34645976	18.055	6.882	356	126736	45118016	18.867	7.087
327	106929	34965783	18.083	6.889	357	127449	45499293	18.894	7 093
328	107584	35287552	18.110	6.896	358	128164	45882712	18.920	7.100
329	108241	35611289	18.138	6.903	359	128881	46268279	18.947	7.107
330	108900	35937000	18.165	6.910	360	129600	46656000	18.973	7.113

THIS TABLE IS TAKEN FROM
"Hamilton's Useful Information for Railway Men,"
Published by D. VAN NOSTRAND,
For the Ramapo Wheel and Foundry Co.

Trucks.

A. M. Gilbert & Co....................................Chicago, Ill......................See Page 180
W. A. Dobinson & Co...................................Buffalo, N. Y.................. " 196

Tube Expanders.

R. Dudgeon...New York.......................See Page 70

Tubing—Brass and Copper.

Ansonia Brass & Copper Co............................New York.......................See Page 122

Turn Buckles.

Providence Tool Co.; H. B. Newhall, Agent....New York.......................See Page 166

Turning Machines.

Wm. Sellers & Co...Philadelphia, Pa.............See Page 64
Wm. B. Bement & Son..................................Philadelphia, Pa............. " 168
See also, Iron Working Machinery and Wood Working Machinery.

Turn Tables.

O. W. Child...New York.......................See Page 262
McNairy & Claflen Manufacturing Co.............Cleveland, Ohio............... " 42
Skinner & Gifford Manufacturing Co..............Dunkirk, N. Y.................. " 140
Snyder Bros...Williamsport, Pa............ " 46
Wm. Sellers & Co...Philadelphia, Pa............. " 64

Twist Drills.

D. Brewer & Co..Philadelphia, Pa.............See Page 246

SQUARES, CUBES, SQUARE ROOTS, AND CUBE ROOTS—*Continued.*

No.	Square.	Cube.	Square Root.	Cube Root.	No.	Square.	Cube.	Square Root.	Cube Root.
361	130321	47045881	19.0	7.120	391	152881	59776471	19.773	7.312
362	131044	47437928	19.026	7.126	392	153664	60236288	19.798	7.318
363	131769	47832147	19.052	7.133	393	154449	60698457	19.824	7.324
364	132496	48228544	19.078	7.140	394	155236	61162984	19.849	7.331
365	133225	48627125	19.104	7.146	395	156025	61629875	19.874	7.337
366	133956	49027896	19.131	7.153	396	156816	62099136	19.899	7.343
367	134689	49430863	19.157	7.159	397	157609	62570773	19.924	7.349
368	135424	49836032	19.183	7.166	398	158404	63044792	19.949	7.355
369	136161	50243409	19.209	7.172	399	159201	63521199	19.974	7.361
370	136900	50653000	19.235	7.179	400	160000	64000000	20.0	7.368
371	137641	51064811	19.261	7.185	401	160801	64481201	20.024	7.374
372	138384	51478848	19.287	7.191	402	161604	64964808	20.049	7.380
373	139129	51895117	19.313	7.198	403	162409	65450827	20.074	7.386
374	139876	52313624	19.339	7.204	404	163216	65939264	20.099	7.392
375	140625	52734375	19.364	7.211	405	164025	66430125	20.124	7.398
376	141376	53157376	19.390	7.217	406	164836	66923416	20.149	7.404
377	142129	53582633	19.416	7.224	407	165649	67419143	20.174	7.410
378	142884	54010152	19.442	7.230	408	166464	67917312	20.199	7.416
379	143641	54439939	19.467	7.236	409	167281	68417929	20.223	7.422
380	144400	54872000	19.493	7.243	410	168100	68921000	20.248	7.428
381	145161	55306341	19.519	7.249	411	168921	69426531	20.273	7.434
382	145924	55742968	19.544	7.255	412	169744	69934528	20.297	7.441
383	146689	56181887	19.570	7.262	413	170569	70444997	20.322	7.447
384	147456	56623104	19.595	7.268	414	171396	70957944	20.346	7.453
385	148225	57066625	19.621	7.274	415	172225	71473375	20.371	7.459
386	148996	57512456	19.646	7.281	416	173056	71991296	20.396	7.465
387	149769	57960603	19.672	7.287	417	173889	72511713	20.420	7.471
388	150544	58411072	19.697	7.293	418	174724	73034632	20.445	7.476
389	151321	58863869	19.723	7.299	419	175561	73560059	20.469	7.482
390	152100	59319000	19.748	7.306	420	176400	74088000	20.493	7.488

THIS TABLE IS TAKEN FROM
"Hamilton's Useful Information for Railway Men,"
Published by D. VAN NOSTRAND,
For the Ramapo Wheel and Foundry Co.

RADLEY & McALISTER MANF'G CO.

MANUFACTURERS OF

GAS, KEROSENE AND SPERM OIL

HEAD LIGHTS.

Also of James Radley's New and Improved

SPARK ARRESTER,

AND THE WELL-KNOWN RADLEY & HUNTER PIPE.

62 John St., New York.

GAS HEAD LIGHT.

Our GAS HEAD LIGHT has now been in use over 4 years, and with entire satisfaction, on 40 of our first-class Railroads, and has demonstrated its superiorty over all others for **Simplicity, Durability, Economy** and great **Brilliancy** of light. Old Kerosene Head Lights altered at small cost. Our **Kerosene** Head Lights are of new patterns. We make 4 sizes· 23 inch, 21 inch, 18 inch and 16 inch.

JAMES RADLEY'S NEW AND IMPROVED SPARK-ARRESTER,

ANSWERS EITHER FOR

WOOD OR COAL BURNING ENGINES.

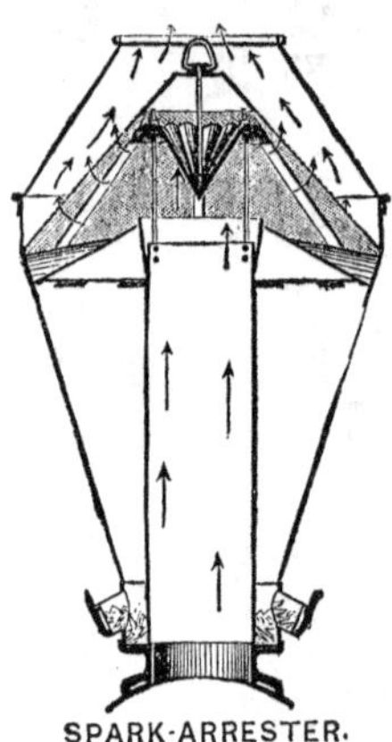

SPARK-ARRESTER.

A large number have been running for a year past on Engines of the *Baltimore and Ohio, Galveston, Houston and Henderson, Galveston, Houston and San Antonio, Houston and Great Northern,* and more recently on the *Louisville and Nashville Railroads.* The Pipe has proved itself the most perfect Spark Arrester known: 1st—As perfectly safe from throwing fire; 2d—The escaping smoke and steam are driven out with great force above the top of the cars and clear of the train, thus keeping the cars free from cinders and smoke; 3d—Its steaming qualities cannot be excelled. By reference to the Cut, it will be seen that the area of the opening of the wire cloth is more than three times that of the opening at the top of the pipe, thus insuring a freer exhaust than in other pipes. For further particulars please send for Circulars.

(Factory, 13 Vine St., Brooklyn, New York.)

Twist Drill Grinders.

D. Brewer & Co.......Philadelphia, Pa.......See Page 246

Twisted Hooks.

D. Brewer & Co.......Philadelphia, Pa.......See Page 246

Type Metal.

Philadelphia Smelting Co.......Philadelphia, Pa.......See Page 134

Tyres.

Jno. A. Griswold & Co.......Troy, N. Y.......See Page 50
L. G. Tillotson & Co.......New York....... " 94
Miller & Smith....... " " .... " 112
Posts & Kalkman.......New York....... " 266
Schenectady Locomotive Works.......Schenectady, N. Y....... " 148
Vose, Dinsmore & Co.......New York....... " 60

Universal Drills.

Wm. Sellers & Co.......Philadelphia, Pa.......See Page 64

Urinals.

A. Carr.......New York.......See Page 244
L. G. Tillotson & Co....... " " .... " 94
Vose, Dinsmore & Co....... " " .... " 60

Vacuum Gauges.

A. Carr.......New York.......See Page 244
Buffalo Steam Gauge Co.......Buffalo, N. Y....... " 170
Geo. W. Richardson & Co.......Troy, N. Y....... " 50
J. G. Knapp Madufacturing Co.......New York....... " 16
McNab & Harlin Manufacturing Co.......New York....... " 220
Tyng & Co....... " " .... " 260
Vose, Dinsmore & Co....... " " .... " 60

SQUARES, CUBES, SQUARE ROOTS, AND CUBE ROOTS—*Continued.*

No.	Square.	Cube.	Square Root.	Cube Root.	No.	Square.	Cube.	Square Root.	Cube Root.
421	177241	74618461	20.518	7.494	451	203401	91733851	21.236	7.668
422	178084	75151448	20.542	7.500	452	204304	92345408	21.260	7.674
423	178929	75686967	20.566	7.506	453	205209	92959677	21.283	7.680
424	179776	76225024	20.591	7.512	454	206116	93576664	21.307	7.685
425	180625	76765625	20.615	7.518	455	207025	94196375	21.330	7.691
426	181476	77308776	20.639	7.524	456	207936	94818816	21.354	7.697
427	182329	77854483	20.663	7.530	457	208849	95443993	21.377	7.702
428	183184	78402752	20.688	7.536	458	209764	96071912	21.400	7.708
429	184041	78953589	20.712	7.541	459	210681	96702579	21.424	7.713
430	184900	79507000	20.736	7.547	460	211600	97336000	21.447	7.719
431	185761	80062991	20.760	7.553	461	212521	97972181	21.470	7.725
432	186624	80621568	20.784	7.559	462	213444	98611128	21.494	7.730
433	187489	81182737	20.808	7.565	463	214369	99252847	21.517	7.736
434	188356	81746504	20.832	7.571	464	215296	99897344	21.540	7.741
435	189225	82312875	20.856	7.576	465	216225	100544625	21.563	7.747
436	190096	82881856	20.880	7.582	466	217156	101194696	21.587	7.752
437	190969	83453453	20.904	7.588	467	218089	101847563	21.610	7.758
438	191844	84027672	20.928	7.594	468	219024	102503232	21.633	7.763
439	192721	84604519	20.952	7.600	469	219961	103161709	21.656	7.769
440	193600	85184000	20.976	7.605	470	220900	103823000	21.679	7.774
441	194481	85766121	21.0	7.611	471	221841	104487111	21.702	7.780
442	195364	86350888	21.023	7.617	472	222784	105154048	21.725	7.785
443	196249	86938307	21.047	7.623	473	223729	105828817	21.748	7.791
444	197136	87528384	21.071	7.628	474	224676	106496424	21.771	7.796
445	198025	88121125	21.095	7.634	475	225625	107171875	21.794	7.802
446	198916	88716536	21.118	7.640	476	226576	107850176	21.817	7.807
447	199809	89314623	21.142	7.646	477	227529	108531333	21.840	7.813
448	200704	89915392	21.166	7.651	478	228484	109215352	21.863	7.818
449	201601	90518849	21.189	7.657	479	229441	109902239	21.886	7.824
450	202500	91125000	21.213	7.663	480	230400	110592000	21.908	7.829

THIS TABLE IS TAKEN FROM
"Hamilton's Useful Information for Railway Men,"
Published by D. VAN NOSTRAND,
For the Ramapo Wheel and Foundry Co.

SQUARES, CUBES, SQUARE ROOTS, AND CUBE ROOTS—*Continued.*

No.	Square.	Cube.	Square Root.	Cube Root.	No.	Square.	Cube.	Square Root.	Cube Root.
481	231361	111284641	21.931	7.835	511	261121	133432831	22.605	7.994
482	232324	111980168	21.954	7.840	512	262144	134217728	22.627	8.0
483	233289	112678587	21.977	7.846	513	263169	135005697	22.649	8.005
484	234256	113379904	22.0	7.851	514	264196	135796744	22.671	8.010
485	235225	114084125	22.022	7.856	515	265225	136590875	22.693	8.015
486	236196	114791256	22.045	7.862	516	266256	137388096	22.715	8.020
487	237169	115501303	22.068	7.867	517	267289	138188413	22.737	8.025
488	238144	116214272	22.090	7.872	518	268324	138991832	22.759	8.031
489	239121	116930169	22.113	7.878	519	269361	139798359	22.781	8.036
490	240100	117649000	22.135	7.883	520	270400	140608000	22.803	8.041
491	241081	118370771	22.158	7.889	521	271441	141420761	22.825	8.046
492	242064	119095488	22.181	7.894	522	272484	142236648	22.847	8.051
493	243049	119823157	22.203	7.899	523	273529	143055667	22.869	8.056
494	244036	120553784	22.226	7.905	524	274576	143877824	22.891	8.062
495	245025	121287375	22.248	7.910	525	275625	144703125	22.912	8.067
496	246016	122023936	22.271	7.915	526	276676	145531576	22.934	8.072
497	247009	122763473	22.293	7.921	527	277729	146363183	22.956	8.077
498	248004	123505992	22.315	7.926	528	278784	147197952	22.978	8.082
499	249001	124251499	22.338	7.931	529	279841	148035889	23.0	8.087
500	250000	125000000	22.360	7.937	530	280900	148877000	23.021	8.092
501	251001	125751501	22.383	7.942	531	281961	149721291	23.043	8.097
502	252004	126506008	22.405	7.947	532	283024	150568768	23.065	8.102
503	253009	127263527	22.427	7.952	533	284089	151419437	23.086	8.107
504	254016	128024064	22.449	7.958	534	285156	152273304	23.108	8.112
505	255025	128787625	22.472	7.963	535	286225	153130375	23.130	8.118
506	256036	129554216	22.494	7.968	536	287296	153990656	23.151	8.128
507	257049	130323843	22.516	7.973	537	288369	154854153	23.173	8 128
508	258064	131096512	22.538	7.979	538	289444	155720872	23.194	8.133
509	259081	131872229	22.561	7.984	539	290521	156590819	23.216	8.138
510	260100	132651000	22.583	7.989	540	291600	157464000	23.237	8.143

THIS TABLE IS TAKEN FROM
"Hamilton's Useful Information for Railway Men,"
Published by D. VAN NOSTRAND,
For the Ramapo Wheel and Foundry Co.

HALL AND SONS'

BUFFALO FIRE BRICK WORKS,

BUFFALO, NEW YORK.

FIRE BRICKS

OF THE BEST QUALITY. FOR ALL PURPOSES,

MANUFACTURED OF THE

Best New Jersey Clays,

FROM OUR OWN MINES.

SPECIAL ATTENTION GIVEN TO ORDERS FOR

LOCOMOTIVE BRICK, ALSO CUPOLAS FOR IRON OR BRASS

N. H. GARDNER & CO.

TANNERS and MANUFACTURERS OF

OAK LEATHER BELTING

AND DEALERS IN

Lace Leather.

Office, 129 Washington St., Tannery, 852 Seneca St.,

BUFFALO, N. Y.

☞ALL GOODS WARRANTED AND SATISFACTION GUARANTEED.

Ventilators.

Vises.

Washers.

SQUARES, CUBES, SQUARE ROOTS, AND CUBE ROOTS—*Continued.*

No.	Square.	Cube.	Square Root.	Cube Root.	No.	Square.	Cube.	Square Root.	Cube Root.
541	292681	158340421	23.259	8 148	571	326041	186169411	23.895	8.296
542	293764	159220088	23.280	8.153	572	327184	187149248	23.916	8.301
543	294849	160103007	23.302	8.158	573	328329	188132517	23 937	8.305
544	295936	160989184	23.323	8.163	574	329476	189119224	23.958	8 310
545	297025	161878625	23.345	8.168	575	330625	190109375	23.979	8.315
546	298116	162771336	23.366	8.173	576	331776	191102976	24.0	8.320
547	299209	163667323	23.388	8.178	577	332929	192100033	24.020	8.325
548	300304	164566592	23 409	8.183	578	334084	193100552	24.041	8.329
549	301401	165469149	23.430	8.188	579	335241	194104539	24.062	8.334
550	302500	166375000	23.452	8.193	580	336400	195112000	24.083	8.339
551	303601	167284151	23.473	8.198	581	337561	196122941	24.103	8.344
552	304704	168196608	23.494	8.203	582	338724	197137368	24.124	8.349
553	305809	169112377	23.515	8.208	583	339889	198155287	24.145	8.353
554	306916	170031464	23.537	8.213	584	341056	199176704	24.166	8.358
555	308025	170953875	23.558	8.217	585	342225	200201625	24.186	8.363
556	309136	171879616	23.579	8.222	586	343396	201230056	24.207	8.368
557	310249	172808693	23.600	8.227	587	344569	202262003	24.228	8.372
558	311364	173741112	23.622	8.232	588	345744	203297472	24.248	8.377
559	312481	174676879	23.643	8.237	589	346921	204336469	24 269	8.382
560	313600	175616000	23.664	8.242	590	348100	205379000	24.289	8.387
561	314721	176558481	23.685	8.247	591	349281	206425071	24.310	8.391
562	315844	177504328	23.706	8.252	592	350464	207474688	24.331	8.396
563	316969	178453547	23.727	8.257	593	351649	208527857	24.351	8.401
564	318096	179406144	23.748	8.262	594	352836	209584584	24.372	8 406
565	319225	180362125	23.769	8.267	595	354025	210644875	24.392	8.410
566	320356	181321496	23.790	8.271	596	355216	211708736	24.413	8.415
567	321489	182284263	23.811	8.276	597	356409	212776173	24.433	8.420
568	322624	183250432	23.832	8.281	598	357604	213847192	24.454	8.424
569	323761	184220009	23.853	8.286	599	358801	214921799	24.474	8.429
570	324900	185193000	23.874	8.291	600	360000	216000000	24.494	8.434

THIS TABLE IS TAKEN FROM
"Hamilton's Useful Information for Railway Men,"
Published by D. VAN NOSTRAND,
For the Ramapo Wheel and Foundry Co.

Washing Machines.

D. Brewer & Co..Philadelphia, Pa.............See Page 246

Waste.

Phelps & Sanger..Cleveland, Ohio..............See Page 120

SAW-MILL.

Gang Saw, 30 square feet of dry oak, or 45 square feet of dry pine, per hour, requires 1 horse-power.

Circular Saw, 2.5 feet in diameter, 270 revolutions per minute, 40 square feet of oak, or 70 of dry spruce, requires 1 horse-power.

300 revolutions per minute. 1.33 square feet of dry pine per minute, kerf 7-32 inch and 6 inches deep, requires the power of 1 horse for the saw alone; and 1 square feet, kerf 1/4 inch and 1 foot in depth, requires a like power.

4.5 feet in diameter, kerf 1/4, and 1 foot in depth, requires 1 horse-power for 1.33 feet per minute.

Oak requires nearly one-half more power than pine.

With a kerf of 1/8 inch, 1 horse-power will saw 2.66 square feet per minute.

The speed of the periphery should be about 50 feet per minute.

SQUARES. CUBES, SQUARE ROOTS, AND CUBE ROOTS—*Continued.*

No.	Square.	Cube.	Square Root.	Cube Root.	No.	Square.	Cube.	Square Root.	Cube Root.
601	361201	217081801	24.515	8.439	631	398161	251239591	25.119	8.577
602	362404	218167208	24.535	8.443	632	399424	252435968	25.139	8.581
603	363609	219256227	24.556	8.448	633	400689	253636137	25.159	8.586
604	364816	220348864	24.576	8.453	634	401956	254840104	25.179	8.590
605	366025	221445125	24.596	8.457	635	403225	256047875	25.199	8.595
606	367236	222545016	24.617	8.462	636	404496	257259456	25.219	8.599
607	368449	223648543	24.637	8.467	637	405769	258474853	25.238	8.604
608	369664	224755712	24.657	8.471	638	407044	259694072	25.258	8.608
609	370881	225866529	24.677	8.476	639	408321	260917119	25.278	8.613
610	372100	226981000	24.698	8.480	640	409600	262144000	25.298	8.617
611	373321	228099131	24.718	8.485	641	410881	263374721	25.317	8.622
612	374544	229220928	24.738	8.490	642	412164	264609288	25.337	8.626
613	375769	230346397	24.758	8.494	643	413449	265847707	25.357	8.631
614	376996	231475544	24.779	8.499	644	414736	267089984	25.377	8.635
615	378225	232608375	24.799	8.504	645	416025	268336125	25.396	8.640
616	379456	233744896	24.819	8.508	646	417316	269585136	25.416	8.644
617	380689	234885113	24.839	8.513	647	418609	270840023	25.436	8.649
618	381924	236029032	24.859	8.517	648	419904	272097792	25.455	8.653
619	383161	237176659	24.879	8.522	649	421201	273359449	25.475	8.657
620	384400	238328000	24.899	8.527	650	422500	274625000	25.495	8.662
621	385641	239483061	24.919	8.531	651	423801	275894451	25.514	8.666
622	386884	240641848	24.939	8.536	652	425104	277167808	25.534	8.671
623	388129	241804367	24.959	8.540	653	426409	278445077	25.553	8.675
624	389376	242970624	24.979	8.545	654	427716	279726264	25.573	8.680
625	390625	244140625	25.0	8.549	655	429025	281011375	25.592	8.684
626	391876	245314376	25.019	8.554	656	430336	282300416	25.612	8.688
627	393129	246491883	25.039	8.558	657	431649	283593393	25.632	8.693
628	394384	247673152	25.059	8.563	658	432964	284890312	25.651	8.697
629	395641	248858189	25.079	8.568	659	434281	286191179	25.670	8.702
630	396900	250047000	25.099	8.572	660	435600	287496000	25.690	8.706

THIS TABLE IS TAKEN FROM
"Hamilton's Useful Information for Railway Men,"
Published by D. VAN NOSTRAND,
For the Ramapo Wheel and Foundry Co.

BOOKS FOR RAILROAD MEN.

SPECIAL NOTICE.—All of the Books from which extracts are made in this work, comprising *Haswell*, *Trautwine*, *Hamilton*, *Haslett*, *&c.*, *&c.*, are for sale by me, and may be ordered by mail.

USEFUL INFORMATION FOR RAILROAD MEN.—Compiled by W. G. HAMILTON, Engineer. Fifth edition, revised and enlarged. 570 pages. Pocket form, Morocco gilt, $2.00.

"The 'Information' comprises some valuable formulæ and rules for the construction of boilers and engines, masonry, properties of steel and iron, and the strength of materials generally."

IRON TRUSS BRIDGES FOR RAILROADS.—The method of calculating strains in Trusses, with a careful comparison of the most prominent Trusses in reference to economy in combination, etc., etc. By Brevet-Colonel W. E. MERRILL, U.S.A., with nine lithographed plates of illustrations. 4to. cloth, $5.00.

A TREATISE ON THE STRENGTH OF BRIDGES AND ROOFS.—With practical applications and examples, for the use of Engineers and Students. By S. H. SHREVE, A.M.C.E. 346 pages, and 89 wood cuts. 8vo., tinted paper, cloth, $5.00.

THE THEORY OF STRAINS IN GIRDERS, and other similar structures, with Observations on the application of Theory to Practice, and Tables of strength and other properties of materials. By BINDON B. STONEY. New edition, revised and enlarged, and complete in one volume, with 5 plates and 123 wood cut illustrations. Royal 8vo., 660 pages, cloth, $15.00,

---

VAN NOSTRAND'S ECLECTIC ENGINEERING MAGAZINE, which commenced its Fifth Year, January, 1873, is admirably adapted to meet the wants, not only of Engineers, but of all who are interested in Scientific subjects. It presents in a convenient form the best articles, (with their illustrations,) selected from European and American Scientific Journals, together with Original Papers from leading scientists of our times.

Issued Monthly, at $5.00 per Annum in Advance.

Single Numbers, 50 Cents.

Persons who desire the Magazine from the beginning, can be supplied with Vols. I to VII, inclusive, neatly bound in cloth, for $20.00; half Turkey morocco, $30.00.

D. VAN NOSTRAND, PUBLISHER,

23 MURRAY ST. & 27 WARREN ST., NEW YORK.

*** Copies of the above Books sent free by mail on receipt of price, and New Catalogue of Scientific Books sent to any address on application.

Water Gauges.

J. G. Knapp Manufacturing Co........................New York......................See Page 16
J. J. Walworth.. Chicago, Ill...................... " 124
McNab & Harlin Manufactnring Co ....New York....................... " 220
Pancoast & Maule..Philadelphia, Pa............. " 26

Water Pipe.

J. T. Ryerson...Chicago, Ill...................... " 180
Morris, Tasker Co..Philadelphia, Pa............. " 36
R. A. Brick & Co.. New York....................... " 48

Wedges.

Hermann Boker & Co......................................New York..............See Pages 152, 154

Wheelbarrows.

Geo. Worthington & Co....................................Cleveland, Ohio ....See Page 100
T. B. Bickerton & Co...Philadelphia, Pa............. " 138

Wheel Presses.

Isaac H. Shearman....................................... Philadelphia, Pa.............See Page 80
Wm. B. Bement & Son................................. " " .... " 168

SQUARES, CUBES, SQUARE ROOTS, AND CUBE ROOTS—*Continued.*

No.	Square.	Cube.	Square Root.	Cube Root.	No.	Square.	Cube.	Square Root.	Cube Root.
661	436921	288804781	25.709	8.710	691	477481	329939371	26.286	8.840
662	438244	290117528	25.729	8.715	692	478864	331373888	26.305	8.845
663	439569	291434247	25.748	8.719	693	480249	332812557	26.324	8.849
664	440896	292754944	25.768	8.724	694	481636	334255384	26.343	8.853
665	442225	294079625	25.787	8.728	695	483025	335702375	26.362	8.857
666	443556	295408296	25.806	8.732	696	484416	337153536	26.381	8.862
667	444889	296740963	25.826	8.737	697	485809	338608873	26.400	8.866
668	446224	298077632	25.845	8.741	698	487204	340068397	26.419	8.870
669	447561	299418309	25.865	8.745	699	488601	341532099	26.438	8.874
670	448900	300763000	25.884	8.750	700	490000	343000000	26.457	8.879
671	450241	302111711	25.903	8.754	701	491401	344472101	26.476	8.883
672	451584	303464448	25.922	8.759	702	492804	345948408	26.495	8.887
673	452929	304821217	25.942	8.763	703	494209	347428927	26.514	8.891
674	454276	306182024	25.961	8.767	704	495616	348913664	26.533	8.895
675	455625	307546875	25.980	8.772	705	497025	350402625	26.551	8.900
676	456976	308915776	26.0	8.776	706	498436	351895816	26.570	8.904
677	458329	310288733	26.019	8.780	707	499849	353393243	26.389	8.908
678	459684	311665752	26.038	8.785	708	501264	354894917	26.608	8.912
679	461041	313046839	26.057	8.789	709	502681	356400829	26.627	8.916
680	462400	314432000	26.076	8.793	710	504100	357911000	26.645	8.921
681	463761	315821241	26.095	8.797	711	505521	359425431	26.664	8.925
682	465124	317214568	26.115	8.802	712	506944	360944128	26.683	8.929
683	466489	318611987	26.134	8.806	713	508369	362467097	26.702	8.933
684	467856	320013504	26.153	8.810	714	509796	363994344	26.720	8.937
685	469225	321419125	26.172	8.815	715	511225	365525875	26.739	8.942
686	470596	322828856	26.191	8.819	716	512656	367061696	26.758	8.946
687	471969	324242703	26.210	8.823	717	514089	368601813	26.776	8.950
688	473344	325660672	26.229	8.828	718	515524	370146232	26.795	8.954
689	474721	327082769	26.248	8.832	719	516961	371694959	26.814	8.958
690	476100	328509000	26.267	8.836	720	518400	373248000	26.832	8.962

THIS TABLE IS TAKEN FROM
"Hamilton's Useful Information for Railway Men,"
Published by D. VAN NOSTRAND,
For the Ramapo Wheel and Foundry Co.

Wheel Borers.

Isaac H. Shearman................................ ....Philadelphia, Pa...........See Page 80

BELTING.

(Hamilton.)

Horse-power of a belt equals velocity in feet per minute, multiplied by the width—the sum divided by 1000.

One inch single belt, moving at 1000 feet per minnte = 1 horse-power.

Double belts about 700 feet per minute, per 1 inch width = 1 horse-power.

For ,couble belts of great length, over large pulleys, allow about 500 feet per minute per 1 inch of width per horse-power.

Power shou'd be communicated through the lower running side of a belt; the the upper side to carry the slack.

Average breaking weights of a belt, $\frac{3}{16} \times 1$ inch wide.

Leather, 530 lbs.; 3-ply rubber, 600 lbs.

The strength of a belt increases directly as its width. The co-efficient of safety for a laced belt is— Leather = $\frac{1}{16}$ breaking weight.
Rubber = $\frac{1}{8}$ " "

SQUARES, CUBES, SQUARE ROOTS, AND CUBE ROOTS—*Continued.*

No.	Square.	Cube.	Square Root.	Cube Root.	No.	Square.	Cube.	Square Root.	Cube Root.
721	519841	374805361	26.851	8.966	751	564001	423564751	27.404	9.089
722	521284	376367048	26.870	8.971	752	565504	425259008	27.422	9.093
723	522729	377933067	26.888	8.975	753	567009	426957777	27.440	9.097
724	524176	379503424	26.907	8.979	754	568516	428661064	27.459	9.101
725	525625	381078125	26.925	8.983	755	570025	430368875	27.477	9.105
726	527076	382657176	26.944	8.987	756	571536	432081216	27.495	9.109
727	528529	384240583	26.962	8.991	757	573049	433798093	27.513	9.113
728	529984	385828352	26.981	8.995	758	574564	435519512	27.531	9.117
729	531441	387420489	27.0	9.0	759	576081	437245479	27.549	9.121
730	532900	389017000	27.018	9 004	760	577600	438976000	27.568	9.125
731	534361	390617891	27.037	9.008	761	579121	440711081	27.586	9.129
732	535824	392223168	27.055	9.012	762	580644	442450728	27.604	9.133
733	537289	393832837	27.073	9.016	763	582169	444194947	27.622	9.137
734	538756	395446904	27.092	9.020	764	583696	445943744	27.640	9.141
735	540225	397065375	27.110	9.024	765	585225	447697125	27.658	9.145
736	541696	398688256	27.129	9.028	766	586756	449455096	27.676	9.149
737	543169	400315553	27.147	9.032	767	588289	451217663	27.694	9.153
738	544644	401947272	27.166	9.036	768	589824	452914832	27.712	9.157
739	546121	403583419	27.184	9.040	769	591361	454756609	27.730	9.161
740	547600	405224000	27.202	9.045	770	592900	456533000	27.748	9.165
741	549081	406869021	27.221	9.049	771	594441	458314011	27.766	9.169
742	550564	408518488	27.239	9.053	772	595984	460099648	27.784	9.173
743	552049	410172407	27.258	9.057	773	597529	461889917	27.802	9.177
744	553536	411830784	27.276	9.061	774	599076	463684824	27.820	9.181
745	555025	413493625	27.294	9.065	775	600625	465484375	27.838	9.185
746	556516	415160936	27.313	9.069	776	602176	467288576	27.856	9.189
747	558009	416832723	27.331	9.073	777	603729	465097433	27.874	9.193
748	559504	418508992	27.349	9.077	778	605284	470910952	27.892	9.197
749	561001	420189749	27.367	9.081	779	606841	472729139	27.910	9.201
750	562500	421875000	27.386	9.085	780	608400	474552000	27.928	9.205

Continued on Page 265

THIS TABLE IS TAKEN FROM
"Hamilton's Useful Information for Railway Men,"
Published by D. VAN NOSTRAND,
For the Ramapo Wheel and Foundry Co.

Wheels.

Bowlers, Maher & Brayton............................Cleveland, Ohio.............See Page 42
Davenport, Fairbairn & Co.............................Erie, Pa.......................... " 188
Greene & Randolph...New York..... .... " 54
Jonas S. Heartt & Co.......................................Troy, N. Y........................ " 134
Ramapo Wheel & Foundry Co.........................Ramapo, N. Y.................. " 256
Wason Manufacturing Co................................Springfield, Mass........ ... " 146

Whistles—Steam.

See Steam Whistles.

White Lead.

Atlantic White Lead Co.—Rob't Colgate & Co. New York...............See Pages 238, 240
Brooklyn White Lead Co............................... " " .... " 268
C. E. Hecht........................... ... .Easton, Pa...................... " 158
F. S. Pease...Buffalo, N. Y................... " 100
Jno. Jewett & Son...New York..................See Pages 72, 74
Jno. W. Masury & Son.................................... " " ....See Page 214
T. B. Bickerton & Co......................................Philadelphia, Pa............. " 138
Wetherill & Bro.... .... " " .... " 52
Wm. Green & Co...Wilmington, Del............ " 248

TONS OF RAILS, JOINTS, ETC., TO ONE MILE SINGLE TRACK.

(*Furnished by O. W. Child, 22 Pine St., New York.*)

Tons of Rails Required per Mile of Single Track.

Weight	Tons.	lbs.	Weight	Tons.	lbs.
Weight 12 lbs. per yard.....	18	1920	Weight 50 lbs. per yard.....	78	1280
" 16 " "	25	320	" 56 " "	88	
" 20 " "	31	960	" 60 " "	94	640
" 25 " "	39	640	" 65 " "	102	320
" 30 " "	47	320	" 70 " "	110	
" 35 " "	55		" 75 " "	117	920
" 40 " "	62	1920	" 80 " "	125	1600
" 45 " "	70	1600			

Number of Fish Joints per Mile.

24 feet Rails.....................	440 Joints.	28 feet Rails.....................	378 Joints.
26 " "	406 "	30 " "	352 "

Ordinary number of Ties per Mile.................................. 2640.
Ordinary Quantity of Spikes, $5\frac{1}{2} \times \frac{9}{16}$ inch per mile...... 5500 lbs.

CULVERTS FOR RAILWAYS.

(*Molesworth*)

	Diameter. ft.	in.	Arch. ft.	in.	Invert. ft.	in.	Sides. ft.	in.	Clear Height Inside. ft.	in.	Cube Yards in 1 Lineal Yard of Culvert.
Brickwork..................	1	6	0	9 ...	Barrel.					...	0.60
	2	0	0	9 ...	Barrel.					...	0.75
	3	0	0	9 ...	0	9 ...	1	2	3	3 ...	1.60
	4	0	1	2 ...	0	9 ...	1	6	4	6 ...	3.30
	5	0	1	2 ...	0	9 ...	1	10	5	6 ...	4.70
	6	0	1	6 ...	0	9 ...	2	3	6	6 ...	6.10
Rubble masonry with hammer-dressed arches.	flat top 1	6	0	6 ...	1	0 ...	1	6	2	6 ...	1.50
	flat top 2	0	0	9 ...	1	0 ...	2	0	3	0 ...	2.00
	3	0	1	0 ...	0	9 ...	2	6	3	3 ...	4.00
	5	0	1	0 ...	1	0 ...	3	0	5	6 ...	6.70
	6	0	1	4 ...	1	0 ...	3	6	5	6 ...	7.60

White Brass.

Du Plaine & Reeves....................................Philadelphia, Pa............See Page 102

White Zinc.

C. E. Hecht..Easton, Pa.....................See Page 158
Jno. W. Masury & Son.................................New York...................... " 214

Window Glass.

D. R. Hobart & Co.......................................New York.......................See Page 126
Leffingwell & Co..Cleveland, Ohio............. " 190
Vose, Dinsmore & Co...................................New York....................... " 60

Wire.

Ansonia Brass & Copper Co..........................New York.......................See Page 122
Atwater, Wheeler & Co.................................New Haven, Conn.......... " 10
Bruce & Cook...New York....................... " 112
Geo. Worthington & Co................................Cleveland, Ohio.............. " 100
Jno. Merry & Co...New York....................... " 48
Marshall Lefferts, Jr...................................." "........................ " 44

TO FIND LENGTH OF BELTING, WHEN CLOSELY ROLLED.

The sum of the diameters of the roll and the eye in inches, multiplied by the number of turns made by the belt, and this product multiplied by the decimal .1309, will equal length of the belt in feet.—*Auchinchloss.*

SQUARES, CUBES, SQUARE ROOTS, AND CUBE ROOTS—*Continued.*

No.	Square.	Cube.	Square Root.	Cube Root.	No.	Square.	Cube.	Square Root.	Cube Root.
781	609961	476379541	27.946	9.209	811	657721	533411731	28.478	9.325
782	611524	478211768	27.964	9.213	812	659344	535387328	28.495	9.329
783	613089	480048687	27.982	9.216	813	660969	537367797	28.513	9.333
784	614656	481890304	28.0	9.220	814	662596	539353144	28.530	9.337
785	616225	483736625	28.017	9.224	815	664225	541343375	28.548	9.340
786	617796	485587656	28.035	9.228	816	665856	543338496	28.565	9.344
787	619369	487443403	28.053	9.232	817	667489	545338513	28.583	9.348
788	620944	489303872	28.071	9.236	818	669124	547343432	28.600	9.352
789	622521	491169069	28.089	9.240	819	670761	549353259	28.618	9.356
790	624100	493039000	28.106	9.244	820	672400	551368000	28.635	9.359
791	625681	494913671	28.124	9.248	821	674041	553387661	28.653	9.363
792	627264	496793088	28.142	9.252	822	675684	555412248	28.670	9.367
793	628849	498677257	28.160	9.256	823	677329	557441767	28.687	9.371
794	630436	500566184	28.178	9.259	824	678976	559476224	28.705	9.375
795	632025	502459875	28.195	9.263	825	680625	561515625	28.722	9.378
796	633616	504358336	28.213	9.267	826	682276	563559976	28.740	9.382
797	635209	506261573	28.231	9.271	827	683929	565609283	28.757	9.386
798	636804	508169592	28.248	9.275	828	685584	567663552	28.774	9 390
799	638401	510082399	28.266	9.279	829	687241	569722789	28.792	9.394
800	640000	512000000	28.284	9.283	830	688900	571787000	28.809	9.397
801	641601	513922401	28.301	9.287	831	690561	573856191	28.827	9.401
802	643204	515849608	28.319	9.290	832	692224	575930368	28.844	9.405
803	644809	517781627	28.337	9.294	833	693889	578009537	28.861	9.409
804	646416	519718464	28.354	9.298	834	695556	580093704	28.879	9.412
805	648025	521660125	28.372	9.302	835	697225	582182875	28.896	9.416
806	649636	523606616	28.390	9.306	836	698896	584277056	28.913	9.420
807	651249	525557943	28.407	9.310	837	700569	586376253	28.930	9.424
808	652864	527514112	28.425	9.314	838	702244	588480472	28.948	9.427
809	654481	529475129	28.442	9.317	839	703921	590589719	28.965	9.431
810	656100	531441000	28.460	9.321	840	705600	592704000	28.982	9.435

THIS TABLE IS TAKEN FROM
"Hamilton's Useful Information for Railway Men,"
Published by D. VAN NOSTRAND,
For the Ramapo Wheel and Foundry Co.

Wire Brushes

Van Tuyl Manufacturing Co..........................New York.....................See Page 242

Wire Cloth.

Radley & McAlister Manufacturing Co............New York..................... See Page 250

Wire Cutters.

Biddle Manufacturing Co................................New York......................See Page 12

Wire Rope.

De Grauw, Aymar & Co.................................New York.........................See Page 112
Speedwell Iron Works.................................. " " .... " 122

Wood Mantels.

Joseph Churchyard....... ....Buffalo, N. Y..................See Page 160

Wood Mouldings.

Joseph Churchyard.......................................Buffalo, N. Y..................See Page 160

Wood Ornaments.

Vose, Dinsmore & Co...................................New York.......................See Page 60

Wood Screws.

Hoopes & Townsend................................ ..... Philadelphia, Pa............See Page 28
Wm. Gilmor of Wm.................. ....Baltimore, Md................. " 248

SQUARES, CUBES, SQUARE ROOTS, AND CUBE ROOTS—*Continued.*

No.	Square.	Cube	Square Root.	Cube Root.	No.	Square.	Cube.	Square Root.	Cube Root.
841	707281	594823321	29.0	9.439	871	758641	660776311	29.512	9.550
842	708964	596947688	29.017	9.442	872	760384	663054848	29.529	9 553
843	710649	599077107	29.034	9.446	873	762129	665338617	29.546	9.557
844	712336	601211584	29.051	9.450	874	763876	667627624	29.563	9.561
845	714025	603351125	29.068	9.454	875	765625	669921875	29.580	9.564
846	715716	605495736	29.086	9.457	876	767376	672221376	29.597	9.568
847	717409	607645423	29.103	9.461	877	769129	674526133	29.614	9.571
848	719104	609800192	29.120	9.465	878	770884	676836152	29.631	9.575
849	720801	611960049	29.137	9.468	879	772641	679151439	29.647	9.579
850	722500	614125000	29.154	9.472	880	774400	681472000	29.664	9.582
851	724201	616295051	29.171	9.476	881	776161	683797841	29.681	9.586
852	725904	618470208	29.189	9.480	882	777924	686128968	29.698	9.590
853	727609	620650477	29.206	9.483	883	779689	688465387	29.715	9.593
854	729316	622835864	29.223	9.487	884	781456	690807104	29.732	9.597
855	731025	625026375	29.240	9.491	885	783225	693154125	29.748	9.600
856	732736	627222016	29.257	9.494	886	784996	695506456	29.765	9.604
857	734449	629422793	29.274	9.498	887	786769	697864103	29.782	9.608
858	736164	631628712	29.291	9.502	888	788544	700227072	29.799	9.611
859	737881	633839779	29.308	9.506	889	790321	702595369	29.816	9.615
860	739600	636056000	29.325	9.509	890	792100	704969000	29 832	9.619
861	741321	638277381	29.342	9 513	891	793881	707347971	29.849	9.622
862	743044	640503928	29.359	9.517	892	795664	709732288	29.866	9.626
863	744769	642735647	29.376	9.520	893	797449	712121957	29.883	9.629
864	746496	644972544	29.393	9 524	894	799236	714516984	29.899	9.633
865	748225	647214625	29.410	9.528	895	801025	716917375	29.916	9.636
866	749956	649461896	29.427	9 531	896	802816	719323136	29.933	9.640
867	751689	651714363	29.444	9.535	897	804609	721734273	29.949	9.644
868	753424	653972032	29.461	9.539	898	806404	724150792	29.966	9.647
869	755161	656234909	29.478	9.542	899	808201	726572699	29.983	9.651
870	756900	658503000	29.495	9.546	900	810000	729000000	30.0	9.654

THIS TABLE IS TAKEN FROM
"Hamilton's Useful Information for Railway Men,"
Published by D. VAN NOSTRAND,
For the Ramapo Wheel and Foundry Co.

BROOKLYN WHITE LEAD CO

(*INCORPORATED 1825.*)

CORRODERS and MANUFACTURERS

OF

PERFECTLY PURE WHITE LEAD.

BEWARE OF SPURIOUS LEADS PUT UP IN IMITATION OF THE BRANDS OF THIS COMPANY.

☞ Every Keg sent out by us is labeled with our ***COPYRIGHT TRADE MARK*** on ***BLUE PAPER,*** and is GUARANTEED to be perfectly pure.

TRADE MARK.

Pure Dry Lead and Litharge,

FOR POTTERS' and INDIA RUBBER MANUFACTURERS' USE.

Office, 89 Maiden Lane, New York.

Wood-Working Machinery.

Buffalo Machinery Agency	Buffalo, N. Y.	See Page 160
C. B. Rogers & Co	New York	" 210
J. A. Fay & Co	Cincinnati, Ohio	" 206
J. T. & R. H. Plass	New York	" 208
Phelps & Sanger	Cleveland, Ohio	" 120
Richards, London & Kelley	Philadelphia, Pa	" 66

Woolen Waste.

T. B. Bickerton & Co	Philadelphia, Pa	See Page 138

Wrecking Derricks.

Skinner & Gifford Manufacturing Co	Dunkirk, N. Y	See Page 140

Wrecking Pumps.

Charles B. Hardick	Brooklyn, N. Y	See Page 84

Wrenches.

Barwick Wrench Co	Boston, Mass	See Page 60
Champlin & Rogers	Chicago, Ill	" 192
Geo. Worthington & Co	Cleveland, Ohio	" 100
J. J. Walworth	Chicago, Ill	" 124
L. G. Tillotson & Co	New York	" 94
Pratt & Co	Buffalo, N. Y	" 132

SQUARES, CUBES, SQUARE ROOTS, AND CUBE ROOTS—*Continued.*

No.	Square.	Cube.	Square Root.	Cube Root.	No.	Square.	Cube.	Square Root.	Cube Root.
901	811801	731432701	30.016	9.658	931	866761	806954491	30.512	9.764
902	813604	733870808	30.033	9.662	932	868624	809557568	30 528	9.767
903	815409	736314327	30.049	9.665	933	870489	812166237	30.545	9.771
904	817216	738763264	30.066	9.669	934	872356	814780504	30.561	9.774
905	819025	741217625	30 083	9.672	935	874225	817400375	30.577	9.778
906	820836	743677416	30.099	9.676	936	876096	820025856	30.594	9.781
907	822649	746142643	30.116	9.679	937	877969	822656953	30.610	9.785
908	824464	748613312	30.133	9.683	938	879844	825293672	30.626	9.788
909	826281	751089429	30.149	9.686	939	881721	827936019	30.643	9.792
910	828100	753571000	30.166	9.690	940	883600	830584000	30.659	9.795
911	829921	756058031	30.182	9.694	941	885481	833237621	30.675	9.799
912	831744	758550528	30.199	9.697	942	887364	835896888	30.692	9.802
913	833569	761048497	30.215	9.701	943	889249	838561807	30.708	9.806
914	835396	763551944	30.232	9.704	944	891136	841232384	30.724	9.809
915	837225	766060875	30.248	9.708	945	893025	843908625	30.740	9.813
916	839056	768575296	30.265	9.711	946	894916	846590536	30.757	9.816
917	840889	771095213	30.282	9.715	947	896809	849278123	30.773	9.820
918	842724	773620632	30.298	9.718	948	898704	851971392	30.789	9.823
919	844561	776151559	30.315	9.722	949	900601	854670349	30.805	9.827
920	846400	778688000	30.331	9.725	950	902500	857375000	30.822	9.830
921	848241	781229961	30.347	9.729	951	904401	860085351	30.838	9.833
922	850084	783777448	30.364	9.732	952	906304	862801408	30.854	9.837
923	851929	786330467	30.380	9.736	953	908209	865523177	30.870	9.840
924	853776	788889024	30.397	9.739	954	910116	868250664	30.886	9.844
925	855625	791453125	30.413	9.743	955	912025	870983875	30.903	9.847
926	857476	794022776	30.430	9.746	956	913936	873722816	30.919	9.851
927	859329	796597983	30.446	9.750	957	915849	876467493	30.935	9.854
928	861184	799178752	30.463	9.754	958	917764	879217912	30.951	9.857
929	863041	801765089	30.479	9.757	959	919681	881974079	30.967	9.861
930	864900	804357000	30.495	9.761	960	921600	884736000	30.983	9.864

THIS TABLE IS TAKEN FROM
"Hamilton's Useful Information for Railway Men,"
Published by D. VAN NOSTRAND,
For the Ramapo Wheel and Foundry Co.

BABCOCK & WILCOX'S *TUBULOUS* SAFETY BOILER.

The best practical evidence of the reliable character of this Boiler is found in the following partial list of our customers:

THE SINGER MANF'G CO., Sewing Machines,	New York,	5 orders,	1560	H.P.
HAVEMEYER & ELDER, Sugar Refiners,	"	2 "	600	"
DECASTRO & DONNER, Sugar Refining Co.,	"	2 "	900	"
F. O. MATTHIESSEN & WIECHERS, Sugar Ref. Co.	"		300	"
HAVEMEYER BROS. & CO., Sugar Refiners,	"	2 "	450	"
HARRISON, HAVEMEYER & CO., Sugar Refiners,	Philadelphia, Pa.		600	"
WAHL BROTHERS, Glue Manufacturers,	Chicago, Ill.,	2 "	300	"
STUDEBAKER BROS'. WAGON CO.,	South Bend, Ind.		300	"
BELCHER SUGAR REFINING CO.,	St. Louis, Mo.		300	"
DAVID TRAINOR & SON, Cotton Mills,	Linwood, Pa.	3 "	425	"
JESSUP & MOORE, Paper Manufacturers,	Wilmington, Del.		200	"
WM. E. HOOPER & SONS, Cotton Mills,	Baltimore, Md.	2 "	325	"
GAMBRILL, SONS & CO., " "	"		150	"
CALVERT SUGAR REFINERY,	"	5 "	1100	"
WOODS, WEEKS & CO., Sugar Refiners,	"	3 "	750	"

And others in all parts of the country. Also all sizes of *Vertical and Horizontal Steam Engines.* ☞ Illustrated Circulars and information sent on application.

BABCOCK & WILCOX, 30 Cortlandt St., N. Y.

New York Safety Steam Power Co.

30 CORTLANDT ST., NEW YORK.

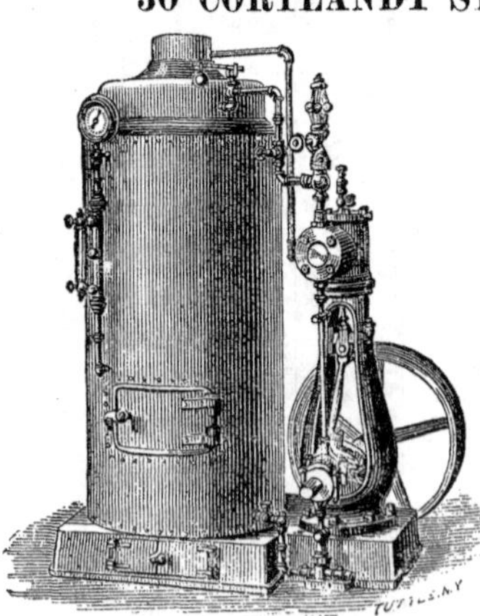

SUPERIOR

Steam Engines and Boilers,

by special machinery and duplication of parts. They are

SAFE,

ECONOMICAL,

EASILY MANAGED,

and not liable to derangement. Their

Combined Engine and Boiler

is peculiarly adapted to all purposes requiring small power. More than 400 Engines, from 2 to 100 horse-power, in use.

☞ *Send for Illustrated Circular.*

Wrought Iron Beams.

Rogers & Co........Chicago, Ill........See Page 144
Vose, Dinsmore & Co........New York........ " 60

Wrought Iron Hardware.

D. Brewer & Co........Philadelphia, Pa........See Page 246

Wrought Iron Pipe.

Crane Bros. Manufacturing Co........Chicago, Ill........See Page 230
Evans, Dalzell & Co........Pittsburgh, Pa........ " 58
Hart, Ball & Hart........Buffalo, N. Y........ " 136
Morris, Tasker & Co........Philadelphia, Pa........ " 36
Pancoast & Maule........ " " .... " 26
Tyng & Co........New York........ " 260
Wm. Graff & Co........Pittsburgh, Pa........ " 174

Wrought Iron Tubes.

Morris, Tasker & Co........Philadelphia, Pa........See Page 36

Zinc.

Bruce & Cook........New York........See Page 112
Jay & Cook........ " " .... " 234
Jno. W. Quincy........ " " .... " 176
Sidney Shepard & Co........Buffalo, N. Y........ " 196

SQUARES, CUBES, SQUARE ROOTS, AND CUBE ROOTS—*Continued.*

No.	Square.	Cube.	Square Root.	Cube Root.	No.	Square.	Cube.	Square Root.	Cube Root.
961	923521	887503681	31.0	9.868	981	962361	944076141	31.320	9.936
962	925444	890277128	31.016	9.871	982	964324	946966168	31.336	9.939
963	927369	893056347	31.032	9.875	983	966289	949862087	31.352	9.943
964	929296	895841344	31.048	9.878	984	968256	952763904	31.368	9.946
965	931225	898632125	31.064	9.881	985	970225	955671625	31.384	9.949
966	933156	901428696	31.080	9.885	986	972196	958585256	31.400	9.953
967	935089	904231063	31.096	9.888	987	974169	961504803	31.416	9.956
968	937024	907039232	31.112	9.892	988	976144	964430272	31.432	9.959
969	938961	909853209	31.128	9.895	989	978121	967361669	31.448	9.963
970	940900	912673000	31.144	9.898	990	980100	970299000	31.464	9.966
971	942841	915498611	31.160	9.902	991	982081	973242271	31.480	9.969
972	944784	918330048	31.176	9.905	992	984064	976191488	31.496	9.973
973	946729	921167317	31.192	9.909	993	986049	979146657	31.511	9.976
974	948676	924010424	31.208	9.912	994	988036	982107784	31.527	9.979
975	950625	926859375	31.225	9.915	995	990025	985074875	31.543	9.983
976	952576	929714176	31.241	9.919	996	992016	988047936	31.559	9.986
977	954529	932574833	31.257	9.922	997	994009	991026973	31.575	9.989
978	956484	935441352	31.272	9.926	998	996004	994011992	31.591	9.993
979	958441	938313739	31.288	9.929	999	998001	997002999	31.606	9.996
980	960400	941192000	31.304	9.932	1000	1000000	1000000000	31.622	10.0

THIS TABLE IS TAKEN FROM
"Hamilton's Useful Information for Railway Men,"
Published by D. VAN NOSTRAND,
For the Ramapo Wheel and Foundry Co.

LENG & OGDEN,

IRON AND STEEL MERCHANTS

4 FLETCHER ST., NEW YORK,

GENERAL AGENTS OF THE

STEAM WATER STATION CO.

MANUFACTURERS OF

LANSDELL'S PATENT STEAM SYPHONS.

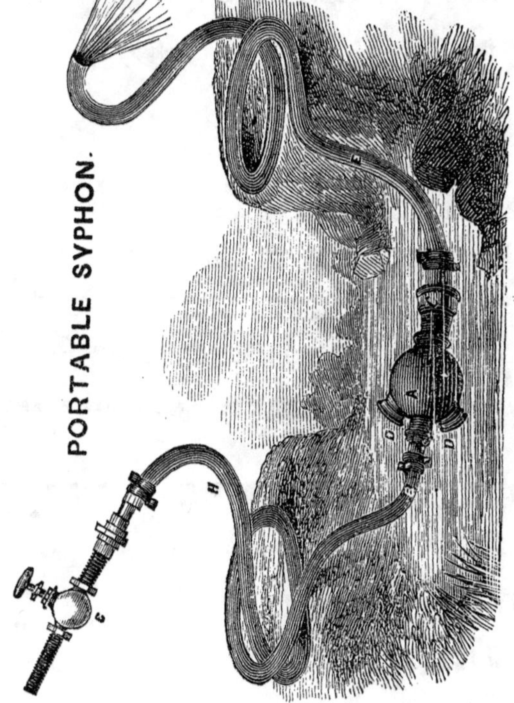

The above shows the mode of applying the Portable Syphon. The valve C is tapped into the dome of the Locomotive boiler; when water is required, the Syphon is thrown into any pond, river, or other water, steam is turned on, the water enters at the suction pipes D, D, and is forced through the hose E into the tender, and will fill an ordinary tender in 6 to 12 minutes.

SOLE AGENTS FOR THE

PATENT WELDLESS COLD-DRAWN STEEL TUBE COMPANY,

AND

CREDENDA STEEL GUN BARRELS, MOULDS, &c.

www.ingramcontent.com/pod-product-compliance
Lightning Source LLC
LaVergne TN
LVHW020220110826
845151LV00003B/775

* 9 7 8 1 4 2 5 5 3 2 2 5 3 *